MILITARY CRYPTANALYSIS. PART III. SIMPLER VARIETIES OF APERIODIC SUBSTITUTION SYSTEMS

CONTENTS

MILITARY CRYPTANALYSIS
Part III

SIMPLER VARIETIES OF APERIODIC SUBSTITUTION SYSTEMS

By
WILLIAM F. FRIEDMAN

This is a quality reproduction of a U.S. Military text, originally published in 1939 — declassified from CONFIDENTIAL in December 1992.

ISBN: 0-89412-196-0 (soft cover)
ISBN: 0-89412-197-9 (library bound)

AEGEAN PARK PRESS
P.O. Box 2837
Laguna Hills, California 92654
(714) 586-8811
FAX (714) 586-8269

Manufactured in the United States of America

SECTION I

INTRODUCTORY

1. Preliminary remarks.—*a.* The text immediately preceding this devoted itself almost exclusively to polyalphabetic substitution systems of the type called repeating-key ciphers. It was seen how a regularity in the employment of a limited number of alphabets results in the manifestation of periodicity or cyclic phenomena in the cryptogram, by means of which the latter may be solved. The difficulty in solution is directly correlated with the type and number of cipher alphabets employed in specific examples.

b. Two procedures suggest themselves for consideration when the student cryptanalyst realizes the foregoing circumstances and thinks of methods to eliminate the weaknesses inherent in this cryptographic system. First, noting that the difficulties in solution increase as the length of the key increases, he may study the effects of employing much longer keys to see if one would be warranted in placing much trust in that method of increasing the security of the messages. Upon second thought, however, remembering that as a general rule the first step in the solution consists in ascertaining the number of alphabets employed, it seems to him that the most logical thing to do would be to use a procedure which will avoid periodicity altogether, will thus eliminate the cyclic phenomena that are normally manifested in cryptograms of a periodic construction, and thus prevent an enemy cryptanalyst from taking even a first step toward solution. In other words, he will investigate the possibilities of *aperiodic* systems first and if the results are unsatisfactory, he will then see what he can do with systems using lengthy keys.

c. Accordingly, the first part of this text will be devoted to an examination of certain of the very simple varieties of aperiodic, polyalphabetic substitution systems; after this, methods of extending or lengthening short mnemonic keys, and systems using lengthy keys will be studied.

2. General remarks upon the nature of cryptographic periodicity.—*a.* When the thoughtful student considers the matter of periodicity in polyalphabetic substitution systems and tries to ascertain its real nature, he notes, with some degree of interest and surprise perhaps that it is composed of *two* fundamental factors, because there are in reality *two* elements involved in its production. He has, of course, become quite familiar with the idea that periodicity necessitates the use of a keying element and that the latter must be employed in a cyclic manner. But he now begins to realize that there is another element involved, the significance of which he has perhaps not fully appreciated, *viz,* that unless the key is applied to constant-length plain-text groups no periodicity will be manifested externally by the cryptogram, despite the repetitive or cyclic use of a constant-length key. This realization is quickly followed by the idea that possibly all periodicity may be avoided or suppressed by either or both of two ways: (1) By using constant-length keying units to encipher variable-length plain-text groupings or (2) by using variable-length keying units to encipher constant-length plain-text groupings.

b. The student at once realizes also that the periodicity exhibited by repeating-key ciphers of the type studied in the preceding text is of a very simple character. There, successive letters of the repetitive key were applied to successive letters of the text. In respect to the employment of the key, the cryptographic or keying process may be said to be *constant* or *fixed* in character. This terminology remains true even if a single keying unit serves to encipher two or more letters at a time, provided only that the groupings of plain-text letters are constant in length. For example, a single key letter may serve to encipher two successive plain-text letters; if the key is repetitive in character and the message is sufficient in length, periodicity will still be manifested by the cryptogram and the latter can be solved by the methods indicated in the preceding text.[1] Naturally, those methods would have to be modified in accordance with the specific type of grouping involved. In this case the factoring process would disclose an apparent key length twice that of the real length. But study of the frequency distributions would soon show that the 1st and 2d distributions were similar, the 3d and 4th, the 5th and 6th, and so on, depending upon the length of the key. The logical step is therefore to combine the distributions in proper pairs and proceed as usual.

c. In all such cases of encipherment by constant-length groupings, the apparent length of the period (as found by applying the factoring process to the cryptogram) is a multiple of the real length and the multiple corresponds to the length of the groupings, that is, the number of plain-text letters enciphered by the same key letter.

d. The point to be noted, however, is that all these cases are still periodic in character, because *both* the keying units and the plain-text groupings are constant in length.

3. Effects of varying the length of the plain-text groupings.—*a.* But now consider the effects of making one or the other of these two elements *variable* in length. Suppose that the plain-text groupings are made variable in length and that the keying units are kept constant in length. Then, even though the key may be cyclic in character and may repeat itself many times in the course of encipherment, external periodicity is suppressed, *unless the law governing the variation in plain-text groupings is itself cyclic in character, and the length of the message is at least two or more times that of the cycle applicable to this variable grouping.*

b. (1) For example, suppose the correspondents agree to use reversed standard cipher alphabets with the key word SIGNAL, to encipher a message, the latter being divided up into groups as shown below:

S	I	G	N	A	L	S	I	G	N	A	L	S	I	G
1	12	123	1234	12345	1	12	123	1234	12345	1	12	123	1234	12345
C	OM	MAN	DING	GENER	A	LF	IRS	TARM	YHASI	S	SU	EDO	RDER	SEFFE
Q	UW	UGT	KFAH	UWNWJ	L	HN	ARQ	NGPU	PGNVF	I	TR	OPE	RFER	OCBBC

N	A	L	S	I	G	N	A	L	S	I	G	N	A	L
1	12	123	1234	12345	1	12	123	1234	12345	1	12	123	1234	12345
C	TI	VET	WENT	YFIRS	T	AT	NOO	NDIR	ECTIN	G	TH	ATT	ELEP	HONES
L	HS	QHS	WOFZ	KDARQ	N	NU	NMM	YIDU	OQZKF	C	NZ	NUU	WPWL	EXYHT

S	I	G	N	A	L	S	I
1	12	123	1234	12345	1	12	123
C	OM	MAS	WITC	HBOAR	D	SC	OMM...
Q	UW	UGO	RFUL	TZMAJ	I	AQ	UWW...

CRYPTOGRAM

QUWUG	TKFAH	UWNWJ	LHNAR	QNGPU	PGNVF	ITROP	ERFER
OCBBC	LHSQH	SWOFZ	KDARQ	NNUNM	MYIDU	OQZKF	CNZNU
UWPWL	EXYHT	QUWUG	ORFUL	TZMAJ	IAQUW	W...	

FIGURE 1.

[1] In this connection, see Section III, *Military Cryptanalysis, Part II*.

(2) The cipher text in this example (Fig. 1) shows a tetragraphic and a pentagraphic repetition. The two occurrences of QUWUG (=COMMA) are separated by an interval of 90 letters; the two occurrences of ARQN (=IRST) by 39 letters. The former repetition (QUWUG), it will be noted, is a true periodic repetition, since the plain-text letters, their grouping, and the key letters are identical. The interval in this case, if counted in terms of letters, is the product of the keying cycle, 6, by the grouping cycle, 15. The latter repetition (ARQN) is not a true periodic repetition in the sense that both cycles have been completed at the same point, as is the case in the former repetition. It is true that the cipher letters ARQN, representing IRST both times, are produced by the same key letters, I and G, but the enciphering points in the grouping cycle are different in the two cases. Repetitions of this type may be termed *partially periodic* repetitions, to distinguish them from those of the *completely periodic* type.

c. When the intervals between the two repetitions noted above are more carefully studied, especially from the point of view of the interacting cycles which brought them about, it will be seen that counting according to *groupings* and not according to single letters, the two pentagraphs QUWUG are separated by an interval of 30 groupings. Or, if one prefers to look at the matter in the light of the keying cycle, the two occurrences of QUWUG are separated by 30 key letters. Since the key is but 6 letters in length, this means that the key has gone through 5 cycles. Thus, the number 30 is the product of the number of letters in the keying cycle (6) by the number of different-length groupings in the grouping cycle (5). The interaction of these two cycles may be conceived of as partaking of the nature of two gears which are in mesh, one driven by the other. One of these gears has 6 teeth, the other 5, and the teeth are numbered. If the two gears are adjusted so that the "number 1 teeth" are adjacent to each other, and the gears are caused to revolve, these two teeth will not come together again until the larger gear has made 5 revolutions and the smaller one 6. During this time, a total of 30 meshings of individual teeth will have occurred. But since one revolution of the smaller gear (=the grouping cycle) represents the encipherment of 15 letters, when translated in terms of letters, the 6 complete revolutions of this gear mean the encipherment of 90 letters. This accounts for the period of 90, when stated in terms of letters.

d. The two occurrences of the other repetition, ARQN, are at an interval of 39 *letters;* but in terms of the number of intervening groupings, the interval is 12, which is obviously two times the length of the keying cycle. In other words, the key has in *this* case passed through 2 cycles.

e. In a long message enciphered according to such a scheme as the foregoing there would be many repetitions of both types discussed above (the completely periodic and the partially periodic) so that the cryptanalyst might encounter some difficulty in his attempts to reach a solution, especially if he had no information as to the basic system. It is to be noted in this connection that if any one of the groupings exceeds say 5, 6, or 7 letters in length, the scheme may give itself away rather easily, since it is clear that *within each grouping the encipherment is strictly monoalphabetic.* Therefore, in the event of groupings of more than 5 or 6 letters, the monoalphabetic equivalents of tell-tale words such as ATTACK, BATTALION, DIVISION, etc., would stand out. The system is most efficacious, therefore, with short groupings.

f. It should also be noted that there is nothing about the scheme which requires a regularity in the grouping cycle such as that embodied in the example. A lengthy grouping cycle such as the one shown below may just as easily be employed, it being guided by a key of its own; for example, the *number* of dots and dashes contained in the International Morse signals for the letters composing the phrase DECLARATION OF INDEPENDENCE might be used. Thus, A (.—) has 2, B (—...) has 4, and so on. Hence:

```
D E C L A R A T I O N O F I N D E P E N D E N C E
3 1 4 4 2 3 2 1 2 3 2 3 4 2 2 3 1 4 1 2 3 1 2 4 1
```

The grouping cycle is 3+1+4+4+2 . . ., or 60 letters in length. Suppose the same phrase is used as an enciphering key for determining the selection of cipher alphabets. Since the phrase contains 25 letters, the complete period of the system would be the least common multiple of 25 and 60 or 300 letters. This system might appear to yield a very high degree of cryptographic security. But the student will see as he progresses that the security is not so high as he may at first glance suppose it to be.

4. Primary and secondary periods; resultant periods.—*a.* It has been noted that the length of the complete period in a system such as the foregoing is the least common multiple of the length of the two component or interacting periods. In a way, therefore, since the *component* periods constitute the *basic* element of the scheme, they may be designated as the *basic* or *primary* periods. These are also *hidden* or *latent* periods. The *apparent* or *patent* period, that is, the complete period, may be designated as the *secondary* or *resultant* period. In certain types of cipher machines there may be more than two primary periods which interact to produce a resultant period; also, there are cases in which the latter may interact with another primary period to produce a tertiary period; and so on. The *final*, or *resultant*, or *apparent* period is the one which is usually ascertained first as a result of the study of the intervals between repetitions. This may or may not be broken down into its component primary periods.

b. Although a solution may often be obtained without breaking down a resultant period into its component primary periods, the reading of many messages pertaining to a widespread system of secret communication is much facilitated when the analysis is pushed to its lowest level, that is, to the point where the final cryptographic scheme has been reduced to its simplest terms. This may involve the discovery of a multiplicity of simple elements which interact in successive cryptographic strata.

Section II

SOLUTION OF SYSTEMS USING CONSTANT-LENGTH KEYING UNITS TO ENCIPHER VARIABLE-LENGTH PLAIN-TEXT GROUPINGS, I

5. Introductory remarks.—*a.* The system described in paragraph 3 above is obviously not to be classified as aperiodic in nature, despite the injection of a variable factor which in that case was based upon irregularity in the length of one of the two elements involved in polyalphabetic substitution. The variable factor was there subject to a law which in itself was periodic in character.

b. To make such a system truly aperiodic in character, by elaborating upon the basic scheme for producing variable-length plain-text groupings, would be possible, but impractical. For example, using the same method as is given in paragraph 3*f* for determining the lengths of the groupings, one might employ the text of a book; and if the latter is longer than the message to be enciphered, the cryptogram would certainly show no periodicity as regards the intervals between repetitions, which would be plentiful. However, as already indicated, such a scheme would not be very practical for regular communication between a large number of correspondents, for reasons which are no doubt apparent. The book would have to be safeguarded as would a code; enciphering and deciphering would be quite slow, cumbersome, and subject to error; and, unless the same key text were used for all messages, methods or indicators would have to be adopted to show exactly where encipherment begins in each message. A simpler method for producing constantly changing, aperiodic plain-text groupings therefore, is to be sought.

6. Aperiodic encipherment produced by groupings according to word lengths.—*a.* The simplest method for producing aperiodic plain-text groupings is one which has doubtless long ago presented itself to the student, *viz.*, encipherment according to the actual word lengths of the message to be enciphered.

b. Although the *average* number of letters composing the words of any alphabetical language is fairly constant, *successive* words comprising plain text vary a great deal in this respect, and this variation is subject to no law.[1] In telegraphic English, for example, the mean length of words is 5.2 letters; the words may contain from 1 to 15 or more letters, but the successive words vary in length in an extremely irregular manner, no matter how long the text may be.

c. As a consequence, the use of word lengths for determining the number of letters to be enciphered by each key letter of a repetitive key commends itself to the inexperienced cryptographer as soon as he comes to understand the way in which repeating-key ciphers are solved. If there is no periodicity in the cryptograms, how can the letters of the cipher text, written in

[1] It is true, of course, that the differences between two writers in respect to the lengths and characters of the words contained in their personal vocabularies are often marked and can be measured. These differences may be subject to certain laws, but the latter are not of the type in which we are interested, being psychological rather than mathematical in character. See Rickert, E., *New Methods for the Study of Literature*, University of Chicago Press, Chicago, 1927.

5-letter groups, be distributed into their respective monoalphabets? And if this very first step is impossible, how can the cryptograms be solved?

7. **Solution when direct standard cipher alphabets are employed.**—*a.* Despite the foregoing rhetorical questions, the solution of this case is really quite simple. It merely involves a modification of the method given in a previous text,[2] wherein solution of a monoalphabetic cipher employing a direct standard alphabet is accomplished by completing the plain-component sequence. There, all the words of the entire message come out on a single generatrix of the completion diagram. In the present case, since the individual, separate words of a message are enciphered by different key letters, *these words will reappear on different generatrices of the diagram.* All the cryptanalyst has to do is to pick them out. He can do this once he has found a good starting point, by using a little imagination and following clues afforded by the context.

b. An example will make the method clear. The following message (note its brevity) has been intercepted:

```
TRECS  YGETI  LUVWV  IKMQI  RXSPJ
SVAGR  XUXPW  VMTUC  SYXGX  VHFFB  LLBHG
```

c. Submitting the message to routine study, the first step is to use normal alphabet strips and try out the possibility of direct standard alphabets having been used. The completion diagram for the first 10 letters of the message is shown in figure 2.

d. Despite the fact that the text does not all reappear on the same generatrix, the solution is a very simple matter because the first three words of the message are easily found: CAN YOU GET. The key letters may be sought in the usual manner and are found to be REA. One may proceed to set up the remaining letters of the message on sliding normal alphabets, or one may assume various keywords such as READ, REAL, REAM, etc., and try to continue the decipherment in that way. The former method is easier. The completed solution is as follows:

R	E	A	D	E	R	S
CAN	YOU	GET	FIRST	REGIMENT	BY	RADIO
TRE	CSY	GET	ILUVW	VIKMQIRX	SP	JSVAG

D	I	G	E	S	T
OUR	PHONE	NOW	OUT	OF	COMMISSION
RXU	XPWVM	TUC	SYX	GX	VHFFBLLBHG

e. Note the key in the foregoing case: It is composed of the successive key letters of the phrase READERS DIGEST.

f. The only difficult part of such a solution is that of making the first step and getting a start on a word. If the words are short it is rather easy to overlook good possibilities and thus spend some time in fruitless searching. However, solution must come; if nothing good appears at the beginning of the message, search should be made in the interior of the cryptogram or at the end.

T	R	E	C	S	Y	G	E	T	I
U	S	F	D	T	Z	H	F	U	J
V	T	G	E	U	A	I	G	V	K
W	U	H	F	V	B	J	H	W	L
X	V	I	G	W	C	K	I	X	M
Y	W	J	H	X	D	L	J	Y	N
Z	X	K	I	Y	E	M	K	Z	O
A	Y	L	J	Z	F	N	L	A	P
B	Z	M	K	A	G	O	M	B	Q
C	A	N	L	B	H	P	N	C	R
D	B	O	M	C	I	Q	O	D	S
E	C	P	N	D	J	R	P	E	T
F	D	Q	O	E	K	S	Q	F	U
G	E	R	P	F	L	T	R	G	V
H	F	S	Q	G	M	U	S	H	W
I	G	T	R	H	N	V	T	I	X
J	H	U	S	I	O	W	U	J	Y
K	I	V	T	J	P	X	V	K	Z
L	J	W	U	K	Q	Y	W	L	A
M	K	X	V	L	R	Z	X	M	B
N	L	Y	W	M	S	A	Y	N	C
O	M	Z	X	N	T	B	Z	O	D
P	N	A	Y	O	U	C	A	P	E
Q	O	B	Z	P	V	D	B	Q	F
R	P	C	A	Q	W	E	C	R	G
S	Q	D	B	R	X	F	D	S	H

FIGURE 2.

[2] *Military Cryptanalysis, Part I,* Par. 20.

8. Solution when reversed standard cipher alphabets are employed.—It should by this time hardly be necessary to indicate that the only change in the procedure set forth in paragraph 7*c*, *d* in the case of reversed standard cipher alphabets is that the letters of the cryptogram must be converted into their plain-component (direct standard) equivalents before the completion sequence is applied to the message.

9. Comments on foregoing cases.—*a*. The foregoing cases are so simple in nature that the detailed treatment accorded them would seem hardly to be warranted at this stage of study. However, they are necessary and valuable as an introduction to the more complicated cases to follow.

b. Throughout this text, whenever encipherment processes are under discussion, the pair of enciphering equations commonly referred to as characterizing the so-called Vigenère method will be understood, unless otherwise indicated. This method involves the pair of enciphering equations $\theta_{i/1}=\theta_{k/2}$; $\theta_{p/1}=\theta_{c/2}$, that is, the index letter, which is usually the initial letter of the plain component, is set opposite the key letter on the cipher component; the plain-text letter to be enciphered is sought on the plain component and its equivalent is the letter opposite it on the cipher component.[3]

c. The solution of messages prepared according to the two preceding methods is particularly easy, for the reason that standard cipher alphabets are employed and these, of course, are derived from *known* components. The significance of this statement should by this time be quite obvious to the student. But what if mixed alphabets are employed, so that one or both of the components upon which the cipher alphabets are based are unknown sequences? The simple procedure of completing the plain component obviously cannot be used. Since the messages are polyalphabetic in character, and since the process of factoring cannot be applied, it would seem that the solution of messages enciphered in different alphabets and according to word lengths would be a rather difficult matter. However, it will soon be made clear that the solution is not nearly so difficult as first impression might lead the student to imagine.

---

[3] See in this connection, *Military Cryptanalysis, Part II*, Section II, and Appendix 1.

SECTION III

SOLUTION OF SYSTEMS USING CONSTANT-LENGTH KEYING UNITS TO ENCIPHER VARIABLE-LENGTH PLAIN-TEXT GROUPINGS, II

10. Solution when the original word lengths are retained in the cryptogram.—*a*. This case will be discussed not because it is encountered in practical military cryptography but because it affords a good introduction to the case in which the original word lengths are no longer in evidence in the cryptogram, the latter appearing in the usual 5-letter groups.

b. Reference is made at this point to the phenomenon called idiomorphism, and its value in connection with the application of the principles of solution by the "probable-word" method, as explained in a previous text.[1] When the original word lengths of a message are retained in the cryptogram, there is no difficulty in searching for and locating idiomorphs and then making comparisons between these idiomorphic sequences in the message and special word patterns set forth in lists maintained for the purpose. For example, in the following message note the underlined groups and study the letters within these groups:

MESSAGE

```
XIXLP  EQVIB  VEFHAPFVT  RT  XWK  PWEWIWRD  XM
NTJCTYZL  OAS  XYQ  ARVVRKFONT  BH  SFJDUUXFP
OUVIGJPF  ULBFZ  RV  DKUKW  ROHROZ
```

IDIOMORPHIC SEQUENCES

```
(1) PWEWIWRD   (2) ARVVRKFONT   (3) SFJDUUXFP
(4) ROHROZ
```

c. Reference to lists of words commonly found in military text and arranged according to their idiomorphic patterns or formulae soon gives suggestions for these cipher groups. Thus:

```
(1) PWEWIWRD       (3) SFJDUUXFP
    DIVISION           ARTILLERY

(2) ARVVRKFONT     (4) ROHROZ
    BATTALIONS         OCLOCK
```

[1] *Military Cryptanalysis, Part I*, Par. 33 *a–d*, inclusive.

d. With these assumed equivalents a reconstruction skeleton or diagram of cipher alphabets (forming a portion of a quadricular table) is established, on the hypothesis that the cipher alphabets have been derived from the sliding of a mixed component against the normal sequence. First it is noted that since $O_p=R_c$ both in the word DIVISION and in the word OCLOCK their cipher equivalents must be in the same alphabet. The reconstruction skeleton is then as follows:

	A	B	C	D	E	F	G	H	I	J	K	L	M	N	O	P	Q	R	S	T	U	V	W	X	Y	Z
Division, o'clock..... (1)			O	P					W		Z	H		D	R				I			E				
Battalion .... (2)	R	A							F			K		N	O				T	V						
Artillery ..... (3)	S				X				D			U						F		J					P	

FIGURE 3*a*.

e. Noting that the interval between O and R in the first and second alphabets is the same, direct symmetry of position is assumed. In a few moments the first alphabet in the skeleton becomes as follows:

	A	B	C	D	E	F	G	H	I	J	K	L	M	N	O	P	Q	R	S	T	U	V	W	X	Y	Z
(1)		N	O	P		S	T	V	W	X	Z	H		D	R	A	U		I			E	F		J	K
(2)	R	A							F			K		N	O				T	V						
(3)	S				X				D			U						F		J					P	

FIGURE 3*b*.

f. The key word upon which the mixed component is based is now not difficult to find: HYDRAULIC.

g. (1) To decipher the entire message, the simplest procedure is to convert the cipher letters into their plain-component equivalents (setting the HYDRAULIC . . . Z sequence against the normal alphabet at any point of coincidence) and then completing the plain-component sequence, as usual. The words of the message will then reappear on different generatrices. The key letters may then be ascertained and the solution completed. Thus, for the first three words, the diagram is as follows:

Plain	A	B	C	D	E	F	G	H	I	J	K	L	M	N	O	P	Q	R	S	T	U	V	W	X	Y	Z
Cipher	H	Y	D	R	A	U	L	I	C	B	E	F	G	J	K	M	N	O	P	Q	S	T	V	W	X	Z

```
X I X L P        E Q V I B        V E F H A P F V T
---------        ---------        -----------------
Y H Y G S        K T W H J        W K L A E S L W V
Z I Z H T        L U X I K        X L M B F T M X W
A J A I U        M V Y J L        Y M N C G U N Y X
B K B J V        N W Z K M        Z N O D H V O Z Y
C L C K W        O X A L N        A O P E I W P A Z
D M D L X        P Y B M O        B P Q F J X Q B A
E N E M Y        Q Z C N P        C Q R G K Y R C B
---------        R A D O Q        D R S H L Z S D C
 Ap=Sc           S B E P R        E S T I M A T E D
                 T C F Q S        -----------------
                 U D G R T             Ap=Pc
                 V E H S U
                 W F I T V
                 X G J U W
                 Y H K V X
                 Z I L W Y
                 A J M X Z
                 B K N Y A
                 C L O Z B
                 D M P A C
                 E N Q B D
                 F O R C E
                 ---------
                  Ap=Uc
```

FIGURE 4.

(2) The key for the message is found to be SUPREME COURT and the complete message is as follows:

SOLUTION

S	U	P	R	E	M	E	C	O	U
ENEMY	FORCE	ESTIMATED	AS	ONE	DIVISION	OF	INFANTRY	AND	TWO
XIXLP	EQVIB	VEFHAPFVT	RT	XWK	PWEWIWRD	XM	NTJCTYZL	OAS	XYQ

R	T	S	U	P	R	E	M
BATTALIONS	OF	ARTILLERY	MARCHING	NORTH	AT	SEVEN	OCLOCK
ARVVRKFONT	BH	SFJDUUXFP	OUVIGJPF	ULBFZ	RV	DKUKW	ROHROZ

h. In case the plain component is the reversed normal sequence, the procedure is no different from the foregoing, except that in the completion diagram the reversed sequence is employed after the cipher letters have been converted into their plain-component equivalents.

i. No doubt the student realizes from his previous work that once the primary mixed component has been recovered the latter becomes a *known* sequence and that the solution of subsequent messages employing the same set of derived alphabets, even though the keys to individual messages are different, then becomes a simple matter.

11. Solution when other types of alphabets are employed.—*a.* The foregoing examples involve the use either of standard cipher alphabets or of mixed cipher alphabets produced by the sliding of a mixed component against the normal sequence. There is, however, nothing about the general cryptographic scheme which prevents the use of other types of derived, interrelated, or secondary mixed alphabets. Cipher alphabets produced by the sliding of a mixed component against itself (either direct or reversed) or by the sliding of two different mixed components are very commonly encountered in these cases.

b. The solution of such cases involves only slight modifications in procedure, namely, those connected with the reconstruction of the primary components. The student should be in a position to employ to good advantage and without difficulty what he has learned about the principles of indirect symmetry of position in the solution of cases of the kind described.

c. The solution of a message prepared with mixed alphabets derived as indicated in subparagraph *b*, may be a difficult matter, depending upon the length of the message in question. It might, of course, be almost impossible if the message is short and there is no background for the application of the probable-word method. But if the message is quite long, or, what is more probable with respect to military communications, should the system be used for regular traffic, so that there are available for study several messages enciphered by the same set of alphabets, then the problem becomes much easier. In addition to the usual steps in solution by the probable-word method, guided by a search for and identification of idiomorphs, there is the help that can be obtained from the use of the phenomena of *isomorphism*, a study of which forms the subject of discussion in the next paragraph.

12. Isomorphism and its importance in cryptanalytics.—*a.* The term idiomorphism is familiar to the student. It designates the phenomena arising from the presence and positions of repeated letters in plain-text words, as a result of which such words may be classified according to their *compositions*, "*patterns*," or *formulae*. The term *isomorphism* (from the Greek "isos" meaning "equal" and "morphe" meaning "form") designates the phenomena arising from the existence of two or more idiomorphs with identical formulae. Two or more sequences which possess identical formulae are said to be *isomorphic*.

b. Isomorphism may exist in plain text or in cipher text. For example, the three words WARRANT, LETTERS, and MISSION are isomorphic. If enciphered monoalphabetically, their cipher equivalents would also be isomorphic. In general, isomorphism is a phenomenon of monoalphabeticity (either plain or cipher); but there are instances wherein it is latent and can be made patent in polyalphabetic cryptograms.

c. In practical cryptanalysis the phenomena of isomorphism afford a constantly astonishing source of clues and aids in solution. The alert cryptanalyst is always on the lookout for situations in which he can take advantage of these phenomena, for they are among the most interesting and most important in cryptanalytics.

13. Illustration of the use of isomorphism.—*a.* Let us consider the case discussed under paragraph 10, wherein a message was enciphered with a set of mixed cipher alphabets derived from sliding the key word-mixed primary component HYDRAULIC . . . XZ against the normal sequence. Suppose the message to be as follows (for simplicity, original word lengths are retained):

CRYPTOGRAM

V C L L K I D V S J D C I O R K D C F S T V I X H M P P F X U E V Z Z
F K N A K F O R A D K O M P I S E C S P P H Q K C L Z K S Q L P R O
J Z W B C X H O Q C F F A O X R O Y X A N O E M D M Z M T S
T Z F V U E A O R S L A U P A D D E R X P N B X A R I G H F X J X I

b. (1) Only a few minutes inspection discloses the following three sets of isomorphs:

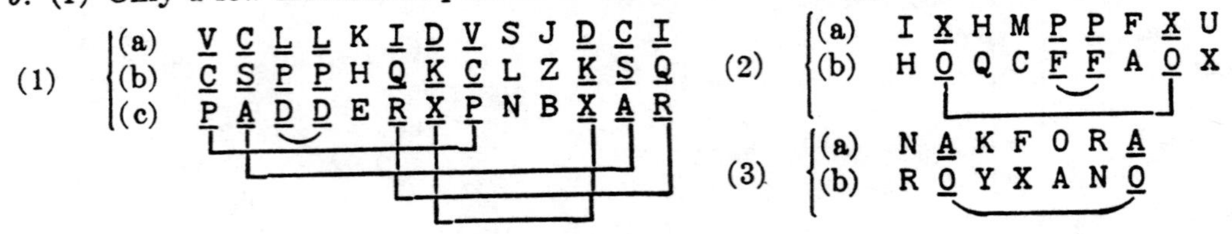

(2) Without stopping to refer to word-pattern lists in an attempt to identify the very striking idiomorphs of the first set, let the student proceed to build up partial sequences of equivalents, as though he were dealing with a case of indirect symmetry of position. Thus:[2]

From isomorphs (1) (a) and (1) (b):

V⇔C; C⇔S; L⇔P; K⇔H; I⇔Q; D⇔K; S⇔L; J⇔Z;

from which the following partial sequences are constructed:

(a) VCSLP (b) DKH (c) IQ (d) JZ

From isomorphs (1) (b) and (1) (c):

C⇔P; S⇔A; P⇔D; H⇔E; Q⇔R; K⇔X; L⇔N; Z⇔B;

from which the following partial sequences are constructed:

(e) CPD (f) SA (g) HE (h) QR (i) KX (j) LN (k) ZB

From isomorphs (1) (a) and (1) (c):

V⇔P; C⇔A; L⇔D; K⇔E; I⇔R; D⇔X; S⇔N; J⇔B;

from which the following partial sequences are constructed:

(l) LDX (m) VP (n) CA (o) KE (p) IR (q) SN (r) JB

Noting that the data from the three isomorphs of this set may be combined (VCSLP and CPD make VCSLP . . D; the latter and LDX make VCSLP . . D . . . X), the following sequences are established:

	1	2	3	4	5	6	7	8	9	10	11	12	13
(1)	V	C	S	L	P	A	N	D	K	H	.	X	E
(2)	I	Q	.	.	R								
(3)	J	Z	.	.	B								

c. (1) The fact that the longest of these chains consists of exactly 13 letters and that no additions can be made from the other two cases of isomorphism, leads to the assumption that a "half-chain" is here disclosed and that the latter represents a decimation of the original primary component at an even interval. Noting the placement of the letters

1	2	3	4	5	6	7	8	9
V	.	S	.	P	.	N	.	K ,

[2] The symbol ⇔ is to be read "is equivalent to."

which gives the sequence the appearance of being the latter half of a keyword-mixed sequence running in the reversed direction, let the half-chain be reversed and extended to 26 places, as follows:

1	2	3	4	5	6	7	8	9	10	11	12	13	14	15	16	17	18	19	20	21	22	23	24	25	26
E		.		K		N		P		S		V		X		H		D		A		L		C	

(2) The data from the two partial chains (JZ..B and IQ..R) may now be used,and the letters inserted into their proper positions. Thus:

1	2	3	4	5	6	7	8	9	10	11	12	13	14	15	16	17	18	19	20	21	22	23	24	25	26
E	.	.	J	K	.	N	.	P	Q	S	.	V	.	X	Z	H	.	D	R	A	.	L	I	C	B

(3) The sequence H . D R A . L I C soon suggests HYDRAULIC as the key word. When the mixed sequence is then developed in full, complete corroboration will be found from the data of isomorphs 2 (a) (b) and 3 (a) (b). Thus:

1	2	3	4	5	6	7	8	9	10	11	12	13	14	15	16	17	18	19	20	21	22	23	24	25	26
H	Y	D	R	A	U	L	I	C	B	E	F	G	J	K	M	N	O	P	Q	S	T	V	W	X	Z

(4) From idiomorphs (2) (a) and (2) (b), the interval between H and I is 7; it is the same for O and X, Q and H, C and M, etc. From idiomorphs (3) (a) and (3) (b) the interval between R and N is 13; it is the same for O and A, Y and K, etc.

d. The message may now be solved quite readily, by the usual process of converting the cipher-text letters into their plain-component equivalents and then completing the plain component sequences. The solution is as follows:

[Key: STRIKE WHILE THE IRON IS . . . (HOT?)]

```
              S                 T         R               I             K
C O M M U N I C A T I O N   W I T H   F I R S T   A R T I L L E R Y   W I L L
V C L L K I D V S J D C I   O R K D   C F S T V   I X H M P P F X U   E V Z Z

E           W             H         I                 L                   E
B E   T H R O U G H   C O R P S   A N D   C O M M U N I C A T I O N   W I T H
F K   N A K F O R A   D K O M P   I S E   C S P P H Q K C L Z K S Q   L P R O

      T                 H               E                 I
S E C O N D   A R T I L L E R Y   T H R O U G H   D I V I S I O N
J Z W B C X   H O Q C F F A O X   R O Y X A N O   E M D M Z M T S

            R             O               N                     I       S
S W I T C H B O A R D   N O   C O M M U N I C A T I O N   A F T E R   T E N
T Z F V U E A O R S L   A U   P A D D E R X P N B X A R   I G H F X   J X I
```

e. (1) In the foregoing illustration the steps are particularly simple because of the following circumstances:

(a) The actual word lengths are shown.

(b) The words are enciphered monoalphabetically by different alphabets belonging to a set of secondary alphabets.

(c) Repetitions of plain-text words, enciphered by different alphabets, produce isomorphs and the lengths of the isomorphs are definitely known as a result of circumstance (a).

(2) Of these facts, the last is of most interest in the present connection. But what if the actual word lengths are not shown; that is, what if the text to be solved is intercepted in the usual 5-letter-group form?

462861 O - 42 - 2

SECTION IV

SOLUTION OF SYSTEMS USING CONSTANT-LENGTH KEYING UNITS TO ENCIPHER VARIABLE-LENGTH PLAIN-TEXT GROUPINGS, III

14. General remarks.—*a*. The cases described thus far are particularly easy to solve because the cryptanalyst has before him the messages in their true or original word lengths. But in military cryptography this is seldom or never the case. The problem is therefore made somewhat more difficult by reason of the fact that there is nothing to indicate definitely the limits of encipherment by successive keyletters. However, the solution merely necessitates more experimentation in this case than in the preceding. The cryptanalyst must take careful note of repetitions which may serve to "block out" or delimit words, and hope that when this is done he will be able to find and identify certain sequences having familiar idiomorphic features or patterns, such as those noted above. If there is plenty of text, repetitions will be sufficient in number to permit of employing this entering wedge.

b. Of course, if any sort of stereotypic phraseology is employed, especially at the beginnings or endings of the messages, the matter of assuming values for sequences of cipher letters is easy, and affords a quick solution. For example, suppose that as a result of previous work it has been found that many messages begin with the expression REFERRING TO YOUR NUMBER . . . . Having several messages for study, the selection of one which begins with such a common idiomorphism as that given by the word REFERRING is a relatively simple matter; and having found the word REFERRING, if with a fair degree of certainty one can add the words TO YOUR NUMBER, the solution is probably well under way.

c. (1) Take the case discussed in paragraph 13, but assume that word lengths are no longer indicated because the message is transmitted in the usual 5-letter groups. The process of ascertaining the exact length of sequences which are isomorphic, or, as the process is briefly termed, "blocking out isomorphs" becomes a more difficult matter and must often rest upon rather tenuous threads of reasoning. For example, take the illustrative message just dealt with and let it be assumed that it was arranged in 5-letter groups.

```
V C L L K   I D V S J   D C I O R   K D C F S   T V I X H   M P P F X
U E V Z Z   F K N A K   F O R A D   K O M P I   S E C S P   P H Q K C
L Z K S Q   L P R O J   Z W B C X   H O Q C F   F A O X R   O Y X A N
O E M D M   Z M T S T   Z F V U E   A O R S L   A U P A D   D E R X P
N B X A R   I G H F X   J X I
```

(2) The detection of isomorphisms now becomes a more difficult matter. There is no special trouble in picking out the following three isomorphic sequences:

```
(1) V C L L K I D V S J D C I
(2) C S P P H Q K C L Z K S Q
(3) P A D D E R X P N B X A R
```

since the first one happens to be at the beginning of the message and its left-hand boundary, or "head," is marked by (or rather, coincides with) the beginning of the message. By a fortunate circumstance, the right-hand boundary, or "tail," can be fixed just as accurately. That the repetition extends as far as indicated above is certain for we have a check on the last column I, Q, R. If an additional column were added, the letters would be O, L, I. Since the second letter has previously appeared while the first and third have not, a contradiction results and the new column may not be included.

If, however, none of the three letters O, L, I had previously appeared, so that there could be no means of getting a check on their correctness, it would not be possible to block out or ascertain the extent of the isomorphism in such a case. All that could be said would be that it seems to include the first 13 letters, *but it might continue further*.

d. (1) However, the difficulty or even the impossibility of blocking out the isomorphs *to their full extent* is not usually a serious matter. After all, the cryptanalyst uses the phenomenon not to identify words but to obtain cryptanalytic data for reconstructing cipher alphabets. For example, how many data are lost when the illustrative message of subparagraph 13*a* is rewritten in 5-letter groups as in subparagraph 14*c* (1)? Suppose the latter form of message be studied for isomorphs:

VCLLK IDVSJ DCIOR KDCFS TVIXH MPPFX UEVZZ
FKNAK FORAD KOMPI SECSP PHQKC LZKSQ LPROJ
ZWBCX HOQCF FAOXR OYXAN OEMDM ZMTST ZFVUE
AORSL AUPAD DERXP NBXAR IGHFX JXI

(2) If the underscored sequences are compared with those in the message in subparagraph 13*a*, it will be found that only a relatively small amount of information has been lost. Certainly not enough to cause any difficulty have been lost in this case, for all the data necessary for the reconstruction of the mixed cipher component came from the first set of isomorphs, and the latter are identical in length in both cases. Only the head and tail letters of the second pair of isomorphic sequences are not included in the underscored sequences in the 5-letter version of the message. The third pair of isomorphic sequences shown in paragraph 13*b* does not appear in the 5-letter version since there is only one repeated letter in this case. In long messages or when there are many short messages, a study of isomorphism will disclose a sufficient number of partial isomorphs to give data usually sufficient for purposes of alphabet reconstruction.

e. It should be noted that there is nothing about the phenomenon of isomorphism which restricts its use to cases in which the cipher alphabets are secondary alphabets resulting from the sliding of a mixed component against the normal. It can be useful in *all* cases of interrelated secondary alphabets no matter what the basis of their derivation may be.

f. In subsequent studies the important role which the phenomenon of isomorphism plays in cryptanalytics will become more apparent. When the traffic is stereotypic in character, even to a slight degree, so that isomorphism may extend over several words or phrases, the phenomenon becomes of highest importance to the cryptanalyst and an extremely valuable tool in his hands.

15. Word separators.—*a.* One of the practical difficulties in employing systems in which the keying process shifts according to word lengths is that in handling such a message the decryptographing clerk is often not certain exactly when the termination of a word has been reached, and thus time is lost by him. For instance, while decryptographing a word such as INFORM the clerk would not know whether he now has the complete word and should shift to the next key letter or not: The word might be INFORMS, INFORMED, INFORMING, INFORMAL, INFOR-

MATION, etc. The past tense of verbs, the plural of nouns, and terminations of various sorts capable of being added to word roots would give rise to difficulties, and the latter would be especially troublesome if the messages contained a few telegraphic errors. Consequently, a scheme which is often adopted to circumvent this source of trouble is to indicate the end of a word by an infrequent letter such as Q or X, and enciphering the letter. In such usage these letters are called *word separators*.

b. When word separators are employed and this fact is once discovered, their presence is of as much aid to the cryptanalyst in his solution as it is to the clerks who are to decryptograph the messages. Sometimes the presence of these word separators, even when enciphered, aids or makes possible the blocking out of isomorphs.

16. Variations and concluding remarks on foregoing systems.—*a.* The systems thus far described are all based upon word-length encipherment using different cipher alphabets. Words are markedly irregular in regard to this feature of their construction, and thus aperiodicity is imparted to such cryptograms. But variations in the method, aimed at making the latter somewhat more secure, are possible. Some of these variations will now be discussed.

b. Instead of enciphering according to natural word lengths, the irregular groupings of the text may be regulated by other agreements. For example, suppose that the numerical value (in the normal sequence) of each key letter be used to control the number of letters enciphered by the successive cipher alphabets. Depending then upon the composition of the key word or key phrase, there would be a varying number of letters enciphered in each alphabet. If the key word were PREPARE, for instance, then the first cipher alphabet would be used for 16 (P=16) letters, the second cipher alphabet, for 18 (=R) letters, and so on. Monoalphabetic encipherment would therefore allow plenty of opportunity for tell-tale word patterns to manifest themselves in the cipher text. Once an entering wedge is found in this manner, solution would be achieved rather rapidly. Of course, all types of cipher alphabets may be employed in this and the somewhat similar schemes described.

c. If the key is short, and the message is long, periodicity will be manifested in the cryptogram, so that it would be possible to ascertain the length of the basic cycle (in this case the length of the key) despite the irregular groupings in encipherment. The determination of the length of the cycle might, however, present difficulties in some cases, since the basic or fundamental period would not be clearly evident because of the presence of repetitions which are not periodic in their origin. For example, suppose the word PREPARE were used as a key, each key letter being employed to encipher a number of letters corresponding to its numerical value in the normal sequence. It is clear that the length of the basic period, in terms of letters, would here be the sum of the numerical values of P (=16)+R (=18) + E (=5), and so on, totalling 79 letters. But because the key itself contains repeated letters and because encipherment by each key letter is monoalphabetic there would be plenty of cases in which the first letter P would encipher the same or part of the same word as the second letter P, producing repetitions in the cryptogram. The same would be true as regards encipherments by the two R's and the two E's in this key word. Consequently, the basic period of 79 would be distorted or masked by aperiodic repetitions, the intervals between which would not be a function of, nor bear any relation to, the length of the key. The student will encounter more cases of this kind, in which a fundamental periodicity is masked or obscured by the presence of cipher-text repetitions not attributable to the fundamental cycle. The experienced cryptanalyst is on the lookout for phenomena of this type, when he finds in a polyalphabetic cipher plenty of repetitions but with no factorable constancy which leads to the disclosure of a short period. He may conclude, then, either that the cryptogram involves several primary periods which interact to produce a long resultant period, or that it involves a fairly long fundamental cycle within which repetitions of a nonperiodic origin are present and obscure the phenomena manifested by repetitions of a periodic origin.

d. (1) A logical extension of the principle of polyalphabetic encipherment of variable-length plain-text groupings is the case in which these plain-text groupings rarely exceed 4 letters, so that a given cipher alphabet is in play for only a very short time, thus breaking up what might otherwise appear as fairly long repetitions in the cipher text. For example, suppose the letters of the alphabet, arranged in their normal-frequency order, were set off into four groups, as follows:

E T R I N	O A S D L C	H F U P M Y G	W V B X K Q J Z
Group 1	Group 2	Group 3	Group 4

(2) Suppose that a letter in group 1 means that one letter will be enciphered; a letter in group 2, that two letters will be enciphered; and so on. Suppose, next, that a rather lengthy phrase were used as a key, for example, PREPARED UNDER THE DIRECTION OF THE CHIEF SIGNAL OFFICER FOR USE WITH ARMY EXTENSION COURSES. Suppose, finally, that each letter of the key were used not only to select the particular cipher alphabet to be used, but also to control the number of letters to be enciphered by the selected alphabet, according to the scheme outlined above. Such an enciphering scheme, using the HYDRAULIC...XZ primary cipher component sliding against the normal plain component, would yield the following groupings:

Grouping	3	1	1	3	2	1	1	2	3	1	2	1	1	1	3	1	2
Key	P	R	E	P	A	R	E	D	U	N	D	E	R	T	H	E	D
Plain	FIR	S	T	DIV	IS	I	O	NW	ILL	A	DV	A	N	C	EAT	F	IV
Cipher	WHB	T	R	THJ	GV	F	X	MX	JNN	N	UW	E	N	W	AHQ	M	EW

Grouping	1	1	1	2	1	1	2	1	2	3	1	3	1	2	3	1	
Key	I	R	E	C	T	I	O	N	O	F	T	H	E	C	H	I	...
Plain	E	F	I	FT	E	E	NA	M	AS	SEC	O	NDD	I	VI	SIO	N	...
Cipher	F	C	P	JY	Z	F	AO	D	OB	RMJ	B	JRR	P	RN	PCK	S	...

(3) Here it will be seen that any tendency for the formation of lengthy repetitions would be counteracted by the short groupings and quick shifting of alphabets. The first time the word DIVISION occurs it is enciphered as THJGVFXM; the second time it occurs it is enciphered as RPRNPCKS. Before DIVISION can be twice enciphered by *exactly* the same sequence of key letters, an interval of at least 140 letters must intervene between the two occurrences of the word (the sum of the values of the letters of the key phrase=140); and then the chances that the key letter P would begin the encipherment of DIV are but one in three. Only one of these three possible encipherments will yield *exactly* the same sequence of cipher equivalents the second time as was obtained the first time. For example, if the text were such as to place two occurrences of the word DIVISION in the positions shown below, their encipherments would be as follows:

3	1	1	3	2	1	1	2	3	1
P	R	E	P	A	R	E	D	U	N
FIR	S	T	DIV	IS	I	O	N	...	.
...	.	.	THJ	GV	F	X	M	...	.

3	1	1	3	2	1	1	2	3	1
P	R	E	P	A	R	E	D	U	N
...	.	.	.DI	VI	S	I	ON	...	.
...	.	.	.TH	ZG	T	P	NM	...	.

Although the word DIVISION, on its second appearance, begins but one letter beyond the place where it begins on its first appearance, the cipher equivalents now agree only in the first two letters, the fourth, and the last letters. Thus:

	D	I	V	I	S	I	O	N
(1)	T	H	J	G	V	F	X	M
(2)	T	H	Z	G	T	P	N	M

e. Attention is directed to the characteristics of the foregoing two encipherments of the same word. When they are superimposed, the first two cipher equivalents are the same in the two encipherments; then there is a single interval where the cipher equivalents are different; the next cipher equivalent is the same; then follow three intervals with dissimilar cipher equivalents; finally, the last cipher equivalent is the same in both cases. The repetitions here extend only to one or two letters; longer repetitions can occur only exceptionally. The two encipherments yield only occasional *coincidences*, that is, places where the cipher letters are identical; moreover, the *distribution* of the coincidences is quite irregular and of an intermittent character.

f. This phenomenon of *intermittent coincidences*, involving coincidences of single letters, pairs of letters, or short sequences (rarely ever exceeding pentagraphs) is one of the characteristics of this general class of polyalphabetic substitution, wherein the cryptograms commonly manifest what appears to be a disturbed or distorted periodicity.

g. From a technical standpoint, the cryptographic principle upon which the foregoing system is based has much merit, but for practical usage it is entirely too slow and too subject to error. However, if the encipherment were mechanized by machinery, and if the enciphering key were quite lengthy, such a system and mechanism becomes of practical importance. Cipher machines for accomplishing this type of substitution will be treated in a subsequent text.

SECTION V

SOLUTION OF SYSTEMS USING VARIABLE-LENGTH KEYING UNITS TO ENCIPHER CONSTANT-LENGTH PLAIN-TEXT GROUPINGS

17. Variable-length groupings of the keying sequence.—The preceding cases deal with simple methods of eliminating or avoiding periodicity by enciphering variable-length groupings of the plain text, using constant-length keying units. In paragraph 2*a*, however, it was pointed out that periodicity can also be suppressed by applying variable-length key groupings to constant-length plain-text groups. One such method consists in *irregularly interrupting* the keying sequence, if the latter is of a limited or fixed length, and recommencing it (from its initial point) after such interruption, so that the keying sequence becomes equivalent to a series of keys of different lengths. Thus, the key phrase BUSINESS MACHINES may be expanded to a series of irregular-length keying sequences, such as BUSI/BUSINE/BU/BUSINESSM/BUSINESSMAC, etc. Various schemes or prearrangements for indicating or determining the interruptions may be adopted. Three methods will be mentioned in the next paragraph.

18. Methods of interrupting a cyclic keying sequence.—*a*. There are many methods of interrupting a keying sequence which is basically cyclic, and which therefore would give rise to periodicity if not interfered with in some way. These methods may, however, be classified into three categories as regards what happens after the interruption occurs:

(1) The keying sequence merely stops and begins again at the initial point of the cycle.

(2) One or more of the elements in the keying sequence may be omitted from time to time irregularly.

(3) The keying sequence irregularly alternates in its direction of progression, with or without omission of some of its elements.

b. These methods may, for clarity, be represented graphically as follows. Suppose the key consists of a cyclic sequence of 10 elements represented symbolically by the series of numbers 1, 2, 3, . . ., 10. Using an asterisk to indicate an interruption, the following may then represent the relation between the letter number of the message and the element number of the keying sequences in the three types mentioned above:

```
    Letter No..........   1 2 3 4    5 6 7 8 9 10    11 12 13    14 15 16 17 18 19 20
(1) Key element No.....   1-2-3-4-*-1-2-3-4-5- 6-*- 1- 2- 3-*- 1- 2- 3- 4- 5- 6- 7-*
    Letter No.......... 21 22 23 24 25 26 27 28 29 30    31 32 33    34 35
    Key element No.....   1- 2- 3- 4- 5- 6- 7- 8- 9-10-*- 1- 2- 3-*- 1- 2- etc.

    Letter No..........   1 2 3    4 5 6 7    8 9    10 11 12    13 14 15 16 17 18 19 20
(2) Key element No.....   1-2-3-*-7-8-9-10-1-2-*- 4- 5- 6-*- 3- 4- 5- 6- 7- 8- 9-10-
    Letter No.......... 21    22 23 24 25 26    27 28 29    30 31 32    33 34 35
    Key element No.....   1-*- 8- 9-10- 1- 2-*- 5- 6- 7-*- 9-10- 1-*- 5- 6- 7- etc.
```

(3) Letter No. 1 2 3 4 5 6 7 8 9 10 11 12 13 14 15 16 17 18 19 20
Key element No. 1–2–3–4–5–*–4–3–*–4–5– 6– 7– 8– 9–10– 1–*–10– 9– 8– 7–*– 8
Letter No. 21 22 23 24 25 26 27 28 29 30 31 32 33 34 35
Key element No. . . . . 9–10– 1– 2– 3–*–2– 1–10– 9– 8–*– 9–10– 1– 2– 3 etc.

As regards the third method, which involves only an alternation in the direction of progression of the keying sequence, if there were no interruptions in the key it would mean merely that a 10-element keying sequence, for example, could be treated as though it were an 18-element sequence and the matter could then be handled as though it were a special form of the second method. But if the principles of the second and third method are combined in one system, the matter may become quite complex.

c. If one *knows* when the interruptions take place in each cycle, then successive *sections* of the basic keying cycle in the three cases may be superimposed. Thus:

METHOD (1)

<table>
<tr><th>Keying element No.</th><th>1</th><th>2</th><th>3</th><th>4</th><th>5</th><th>6</th><th>7</th><th>8</th><th>9</th><th>10</th></tr>
<tr><td>Letter No.</td><td>1</td><td>2</td><td>3</td><td>4 |</td><td></td><td></td><td></td><td></td><td></td><td></td></tr>
<tr><td>Letter No.</td><td>5</td><td>6</td><td>7</td><td>8</td><td>9</td><td>10 |</td><td></td><td></td><td></td><td></td></tr>
<tr><td>Letter No.</td><td>11</td><td>12</td><td>13 |</td><td></td><td></td><td></td><td></td><td></td><td></td><td></td></tr>
<tr><td>Letter No.</td><td>14</td><td>15</td><td>16</td><td>17</td><td>18</td><td>19</td><td>20 |</td><td></td><td></td><td></td></tr>
<tr><td>Letter No.</td><td>21</td><td>22</td><td>23</td><td>24</td><td>25</td><td>26</td><td>27</td><td>28</td><td>29</td><td>30 |</td></tr>
<tr><td>Letter No.</td><td>31</td><td>32</td><td>33 |</td><td></td><td></td><td></td><td></td><td></td><td></td><td></td></tr>
<tr><td>Letter No.</td><td>34</td><td>35</td><td>etc.</td><td></td><td></td><td></td><td></td><td></td><td></td><td></td></tr>
</table>

METHOD (2)

<table>
<tr><th>Keying element No.</th><th>1</th><th>2</th><th>3</th><th>4</th><th>5</th><th>6</th><th>7</th><th>8</th><th>9</th><th>10</th></tr>
<tr><td>Letter No.</td><td>1</td><td>2</td><td>3 |</td><td>—</td><td>—</td><td>—</td><td>4</td><td>5</td><td>6</td><td>7</td></tr>
<tr><td>Letter No.</td><td>8</td><td>9 |</td><td>—</td><td>10</td><td>11</td><td>12 |</td><td></td><td></td><td></td><td></td></tr>
<tr><td>Letter No.</td><td>—</td><td>—</td><td>13</td><td>14</td><td>15</td><td>16</td><td>17</td><td>18</td><td>19</td><td>20</td></tr>
<tr><td>Letter No.</td><td>21 |</td><td>—</td><td>—</td><td>—</td><td>—</td><td>—</td><td>—</td><td>22</td><td>23</td><td>24</td></tr>
<tr><td>Letter No.</td><td>25</td><td>26 |</td><td>—</td><td>—</td><td>27</td><td>28</td><td>29 |</td><td>—</td><td>30</td><td>31</td></tr>
<tr><td>Letter No.</td><td>32 |</td><td>—</td><td>—</td><td>—</td><td>33</td><td>34</td><td>35</td><td>etc.</td><td></td><td></td></tr>
</table>

METHOD (3)

<table>
<tr><th>Keying element No.</th><th>1</th><th>2</th><th>3</th><th>4</th><th>5</th><th>6</th><th>7</th><th>8</th><th>9</th><th>10</th></tr>
<tr><td>Letter No.</td><td>1</td><td>2</td><td>3</td><td>4</td><td>5 |</td><td>—</td><td>—</td><td>—</td><td>—</td><td>—</td></tr>
<tr><td>Letter No.</td><td>—</td><td>— |</td><td>7</td><td>6</td><td>—</td><td>—</td><td>—</td><td>—</td><td>—</td><td>—</td></tr>
<tr><td>Letter No.</td><td>—</td><td>—</td><td>—</td><td>8</td><td>9</td><td>10</td><td>11</td><td>12</td><td>13</td><td>14</td></tr>
<tr><td>Letter No.</td><td>15 |</td><td>—</td><td>—</td><td>—</td><td>—</td><td>— |</td><td>19</td><td>18</td><td>17</td><td>16</td></tr>
<tr><td>Letter No.</td><td>23</td><td>24</td><td>25 |</td><td>—</td><td>—</td><td>—</td><td>—</td><td>20</td><td>21</td><td>22</td></tr>
<tr><td>Letter No.</td><td>27</td><td>26</td><td>—</td><td>—</td><td>—</td><td>—</td><td>— |</td><td>30</td><td>29</td><td>28</td></tr>
<tr><td>Letter No.</td><td>33</td><td>34</td><td>35</td><td>etc.</td><td></td><td></td><td></td><td></td><td>31</td><td>32</td></tr>
</table>

Obviously if one does not know when or how the interruptions take place, then the successive sections of keying elements cannot be superimposed as indicated above.

d. The interruption of the cyclic keying sequence usually takes place according to some prearranged plan, and the three basic methods of interruption will be taken up in turn, using a short mnemonic key as an example.

e. Suppose the correspondents agree that the interruption in the keying sequence will take place after the occurrence of a specified letter called an *interruptor*,[1] which may be a letter of the plain text, or one of the cipher text, as agreed upon in advance. Then, since in either case there is nothing fixed about the time the interruption will occur—it will take place at no fixed intervals—not only does the interruption become quite irregular, following no pattern, but also the method never reverts to one having periodicity. Methods of this type will now be discussed in detail.

19. Interruptor is a plain-text letter.—*a.* Suppose the correspondents agree that the interruption in the key will take place immediately after a previously agreed-upon letter, say R, occurs in the plain text. The key would then be interrupted as shown in the following example (using the mnemonic key BUSINESS MACHINES and the HYDRAULIC . . . XZ sequence):

Key	B U S I N E S S M A C H I	B U S	B U S I	B U S I N E
Plain	A M M U N I T I O N F O R	F I R	S T A R	T I L L E R
Cipher	B O L Y R P J D R O J K X	K J F	Y X S X	D J U P S Y

Key	B U S I N E S S M A C H I N E S B U	B U S I N E S S M A C H I
Plain	Y W I L L B E L O A D E D A F T E R	A M M U N I T I O N F O R
Cipher	I Y D P Y F X U R A F A E N M J J V	B O L Y R P J D R O J K X

Key	B U S I	B U S	B U S I N E	B U S I N
Plain	T H I R	D A R	T I L L E R	Y . . . .
Cipher	D G D X	G U F	D J U P S Y	I . . . .

CRYPTOGRAM

B O L Y R P J D R O J K X K J F Y X S X D J U P S Y I Y D P
Y F X U R A F A E N M J J V B O L Y R P J D R O J K X D G D
X G U F D J U P S Y I X X X X

b. Instead of employing an ordinary plain-text letter as the interruptor, one might reserve the letter J for this purpose (and use the letter I whenever this letter appears as part of a plain-text word). This is a quite simple variation of the basic method. The letter J acts merely as though it were a plain-text letter, except that in this case it also serves as the interruptor. The interruptor is then inserted *at random*, at the whim of the enciphering clerk. Thus:

Key	BUSINESSMAC	BUSINESSM	BUSINESSMACHINESBUSIN
Plain	TROOPSWILLJ	BEHALTEDJ	ATROADIUNCTIONFIVESIX

c. It is obvious that repetitions would be plentiful in cryptograms of this construction, regardless of whether a letter of high, medium, or low frequency is selected as the signal for key interruption. If a letter of high frequency is chosen, repetitions will occur quite often, not only because that letter will certainly be a part of many common words, but also because it will be followed by words that are frequently repeated; and since the key starts again with each such interruption, these frequently repeated words will be enciphered by the same sequence of alphabets. This is the case in the first of the two foregoing examples. It is clear, for instance, that every time the word ARTILLERY appears in the cryptogram the cipher equivalents of TILLERY must be the same. If the interruptor letter were A_p instead of R_p, the repetition

[1] Also called at times an "influence" letter because it influences or modifies normal procedure. In some cases no influence or interruptor letter is used, the interruption or break in the keying sequence occurring after a previously-agreed-upon number of letters has been enciphered.

would include the cipher equivalents of RTILLERY; if it were T_p, ILLERY, and so on. On the other hand, if a letter of low frequency were selected as the interruptor letter, then the encipherment would tend to approximate that of normal repeating-key substitution, and repetitions would be plentiful on that basis alone.

d. Of course, the lengths of the intervals between the repetitions, in any of the foregoing cases, would be irregular, so that periodicity would not be manifested. The student may inquire, therefore, how one would proceed to solve such messages, for it is obvious that an attempt to allocate the letters of a single message into separate monoalphabetic distributions cannot be successful unless the exact locations of the interruptions are known—and they do not become known to the cryptanalyst until he has solved the message, or at least a part of it. Thus it would appear as though the would-be solver is here confronted with a more or less insoluble dilemma. This sort of reasoning, however, makes more of an appeal to the novice in cryptography than to the experienced cryptanalyst, who specializes in methods of solving cryptographic dilemmas.

e. (1) The problem here will be attacked upon the usual two hypotheses, and the easier one will be discussed first. Suppose the system has been in use for some time, that an original solution has been reached by means to be discussed under the second hypothesis, and that the cipher alphabets are known. There remains unknown only the specific key to messages. Examining whatever repetitions are found, an attack is made on the basis of searching for a probable word. Thus, taking the illustrative message in subparagraph *a*, suppose the presence of the word ARTILLERY is suspected. Attempts are made to locate this word, basing the search upon the construction of an intelligible key. Beginning with the very first letter of the message, the word ARTILLERY is juxtaposed against the cipher text, and the key letters ascertained, using the known alphabets, which we will assume in this case are based upon the HYDRAULIC . . . XZ sequence sliding against the normal. Thus:

Cipher.................. B O L Y R P J D R
Plain.................... A R T I L L E R Y
"Key".................. B H J Q P I B F U

(2) Since this "key" is certainly not intelligible text, the assumed word is moved one letter to the right and the test repeated, and so on until the following place in the test is reached:

Cipher.................. S X D J U P S Y I
Plain.................... A R T I L L E R Y
Key..................... S I B U S I N E B

(3) The sequence BUSINE suggests BUSINESS; moreover, it is noted that the key is interrupted both times by the letter R_p. Now the key may be applied to the beginning of the message, to see if the whole key or only a portion of it has been recovered. Thus:

Key..................... B U S I N E S S B U S
Cipher.................. B O L Y R P J D R O J
Plain.................... A M M U N I T I U M T

(4) It is obvious that BUSINESS is only a part of the key. But the deciphered sequence certainly seems to be the word AMMUNITION. When this is tried, the key is extended to BUSINESS MA . . . . Enough has been shown to clarify the procedure.

f. The foregoing solution is predicated upon the hypothesis that the cipher alphabets are known. But what if this is not the case? What of the steps necessary to arrive at the *first* solution, before even the presence of an interruptor is suspected? The answer to this question leads to the presentation of a method of attack which is one of the most important and powerful means the cryptanalyst has at his command for unraveling many knotty problems. It is called *solution by superimposition*, and warrants detailed treatment.

20. Solution by superimposition.— *a. Basic principles.* — (1) In solving an ordinary repeating-key cipher the first step, that of ascertaining the length of the period, is of no significance in itself. It merely paves the way for and makes possible the second step, which consists in allocating the letters of the cryptogram into individual monoalphabetic distributions. The third step then consists in solving these distributions. Usually, the text of the message is transcribed into its periods and is written out in successive lines corresponding in length with that of the period. The diagram then consists of a series of columns of letters and the letters in each column belong to the same monoalphabet. Another way of looking at the matter is to conceive of the text as having thus been transcribed into *superimposed periods:* in such case the letters in each column have undergone the same kind of treatment by the same elements (plain and cipher components of the cipher alphabet).

(2) Suppose, however, that the repetitive key is very long and that the message is short, so that there are only a very few cycles in the text. Then the solution of the message becomes difficult, if not impossible, because there is not a sufficient number of superimposable periods to yield monoalphabetic distributions which can be solved by frequency principles. But suppose also that there are many short cryptograms all enciphered by the same key. Then it is clear that if these messages are superimposed:

(a) The letters in the respective columns will all belong to individual alphabets; and

(b) If there is a sufficient number of such superimposable messages (say 25–30, for English), then the frequency distributions applicable to the successive *columns* of text can be solved—*without knowing the length of the key*. In other words, any difficulties that may have arisen on account of failure or inability to ascertain the length of the period have been circumvented. The second step in normal solution is thus "by-passed."

(3) Furthermore, and this is a very important point, in case an extremely long key is employed and a series of messages beginning at different initial points are enciphered by such a key, this method of solution by superimposition can be employed, provided the messages can be superimposed correctly, that is, so that the letters which fall in one column really belong to one cipher alphabet. Just how this can be done will be demonstrated in subsequent paragraphs, but a clue has already been given in paragraph 18*c*. At this point, however, a simple illustration of the method will be given, using the substitution system discussed in paragraph 19.

b. Example.—(1) A set of 35 messages has been intercepted on the same day. Presumably they are all in the same key, and the presence of repetitions between messages corroborates this assumption. But the intervals between repetitions within the same message do not show any common factor and the messages appear to be aperiodic in nature. The probable-word method has been applied, using standard alphabets, with no success. The messages are then superimposed (Fig. 5); the frequency distributions for the first 10 columns are as shown in Figure 6.

1 Z C T P Z W Z P E P Z Q X
2 W T E Q M X Z S Y S P R C
3 T C R W C X T B H H
4 E F K C S Z R I H A
5 Y A N C I H Z N U W
6 V Z I E T I R R G X
7 H C Q I C K G U O N
8 Z C F C L X R K Q W
9 H W W P T E W C I M J S
10 E P D O Z C L I K S J
11 W T S S Q Z P Z I E T
12 Z C G G Y F C S B G
13 C W Z A O O E M H W T P
14 C I Y G I F B D T V X
15 E A Q D R D N S R C A P D T
16 Y F W C Q Q B Z C W C
17 W T E Z Q S K U H C
18 Z C V X Q Z K Z Y D W L K
19 A F E O J T D T I T
20 K P V F Q W P K T E V
21 Z A B G R T X P U Q X
22 Y H E O C U H M D T
23 C L C P Z I K O T H
24 A F L W W Z Q M D T
25 Z C W A P M B S A W L
26 H F L M H R Z N A P E C E
27 C L Z G E M K Z T O
28 T P Y F K O T I Z U H
29 Z C C P S N E O P H D Y L
30 C I Y G I F T S Y T L E
31 Y T S V W V D G H P G U Z
32 N O C A I F B J B L G H Y
33 Z X X F L F E G J L
34 Z C T M M B Z J O O
35 H C Q I W S Y S B P H C Z V

FIGURE 5.

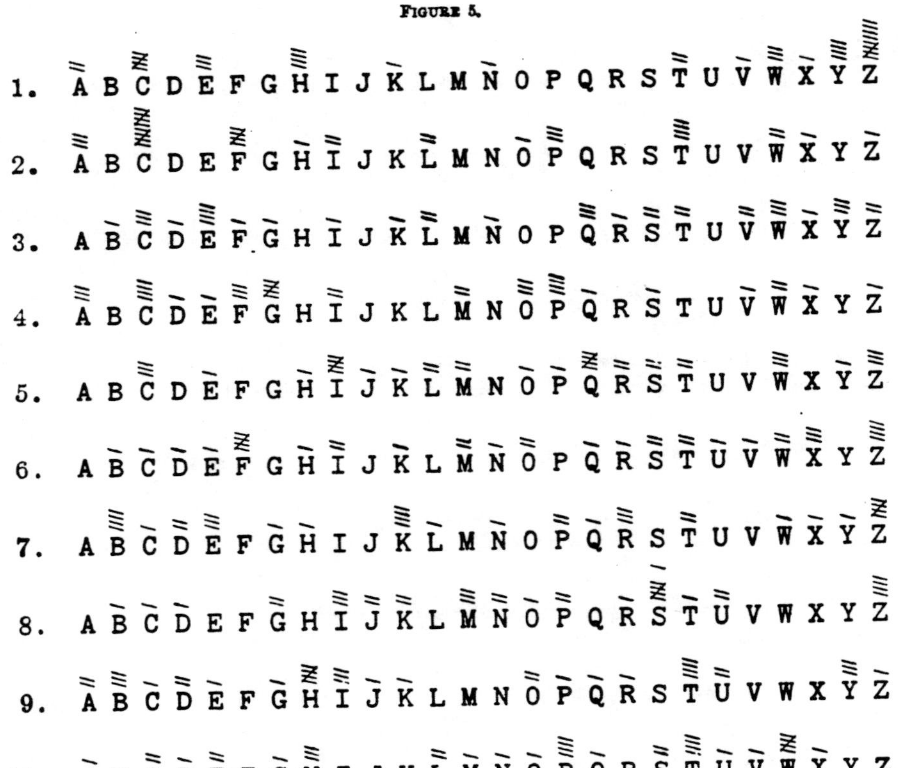

1. A B C D E F G H I J K L M N O P Q R S T U V W X Y Z

2. A B C D E F G H I J K L M N O P Q R S T U V W X Y Z

3. A B C D E F G H I J K L M N O P Q R S T U V W X Y Z

4. A B C D E F G H I J K L M N O P Q R S T U V W X Y Z

5. A B C D E F G H I J K L M N O P Q R S T U V W X Y Z

6. A B C D E F G H I J K L M N O P Q R S T U V W X Y Z

7. A B C D E F G H I J K L M N O P Q R S T U V W X Y Z

8. A B C D E F G H I J K L M N O P Q R S T U V W X Y Z

9. A B C D E F G H I J K L M N O P Q R S T U V W X Y Z

10. A B C D E F G H I J K L M N O P Q R S T U V W X Y Z

FIGURE 6.

(2) The 1st and 2d distributions are certainly monoalphabetic. There are very marked crests and troughs, and the number of blanks (14) is more than satisfactory in both cases. (Let the student at this point refer to Par. 14 and Chart 5 of Military Cryptanalysis, Part I.) But the 3d, 4th, and remaining distributions appear no longer to be monoalphabetic. Note particularly the distribution for the 6th column. From this fact the conclusion is drawn that some disturbance in periodicity has been introduced in the cryptograms. In other words, although they all start out with the same alphabet, some sort of interruption takes place so as to suppress periodicity.

(3) However, a start on solution may be made by attacking the first two distributions, frequency studies being aided by considerations based upon probable words. In this case, since the text comprises only the beginnings of messages, assumptions for probable words are more easily made than when words are sought in the interiors of messages. Such common introductory words as REQUEST, REFER, ENEMY, WHAT, WHEN, IN, SEND, etc., are good ones to assume. Furthermore, high-frequency digraphs used as the initial digraphs of common words will, of course, manifest themselves in the first two columns. The greatest aid in this process is, as usual, a familiarity with the "word habits" of the enemy.

(4) Let the student try to solve the messages. In so doing he will more or less quickly find the cause of the rapid falling off in monoalphabeticity as the columns progress to the right from the initial point of the messages.

21. Interruptor is a cipher-text letter.—*a.* In the preceding case a plain-text letter serves as the interruptor. But now suppose the correspondents agree that the interruption in the key will take place immediately after a previously-agreed-upon letter, say Q, occurs in the cipher text. The key would then be interrupted as shown in the following example:

```
Key.............. B U S I N E S S M A C H I N E S B U S I N E S S M|
Plain............ A M M U N I T I O N F O R F I R S T A R T I L L E|
Cipher........... B O L Y R P J D R O J K X T P F Y X S X B P U U Q|

Key.............. B U S I N E S S M A C H I N|B U S I N E S S M A C H|B U|
Plain............ R Y W I L L B E L O A D E D|A F T E R A M M U N I T|I O|
Cipher........... H R N M Y T T X H P C R F Q|B E J F I E L L B O N Q|O Q|

Key.............. B U S I N E S S M A C H|B U S I N E
Plain............ N F O R T H I R D A R T|I L L E R Y
Cipher........... V E C X B O D F P A Z Q|O N U F I C
```

CRYPTOGRAM

```
B O L Y R   P J D R O   J K X T P   F Y X S X   B P U U Q   H R N M Y
T T X H P   C R F Q B   E J F I E   L L B O N   Q O Q V E   C X B O D
F P A Z Q   O N U F I   C X X X X
```

b. In the foregoing example, there are no significant repetitions. Such as do occur comprise only digraphs, one of which is purely accidental. But the absence of significant, long repetitions is itself purely accidental; for had the interruptor letter been a letter other than Q_c, then the phrase AMMUNITION FOR, which occurs twice, might have been enciphered identically both times. If a short key is employed, repetitions may be plentiful. For example, note the following, in which S_c is the interruptor letter:

Key	B A N D S B A N D S B A N D S B A N D S B A N	B A N D S B A N D S B
Plain	F R O M F O U R F I V E T O F O U R F I F T E	E N A M B A R R A G E
Cipher	K T A K Z W X I I D A C B N Z W X I I D K W S	J O N K T B T I D H J

c. This last example gives a clue to one method of attacking this type of system. There will be repetitions within short sections, and the interval between them will sometimes permit of ascertaining the length of the key. In such short sections, the letters which intervene between the repeated sequences may be eliminated as possible interruptor letters. Thus, the letters A, C, B, and N may be eliminated, in the foregoing example, as interruptor letters. By extension of this principle to the letters intervening between other repetitions, one may more or less quickly ascertain what letter serves as the interruptor.

d. Once the interruptor letter has been found, the next step is to break up the message into "uninterrupted" sequences and then attempt a solution by superimposition. The principles explained in paragraph 20 need only be modified in minor respects. In the first place, in this case the columns of text formed by the superimposition of uninterrupted sequences will be purely monoalphabetic, whereas in the case of the example in paragraph 20, only the very first column is purely monoalphabetic, the monoalphabeticity falling off very rapidly with the 2d, 3d, . . . columns. Hence, in this case the analysis of the individual alphabets should be an easier task. But this would be counterbalanced by the fact that whereas in the former case the cryptanalyst is dealing with the initial words of messages, in this case he is dealing with interior portions of the text and has no way of knowing where a word begins. The latter remarks naturally do not apply to the case where a whole set of messages in this system, all in the same key, can be subjected to simultaneous study. In such a case the cryptanalyst would also have the initial words to work upon.

22. Concluding remarks.—*a*. The preceding two paragraphs both deal with the first and simplest of the three basic cases referred to under paragraph 18. The second of those cases involves considerably more work in solution for the reason that when the interruption takes place and the keying sequence recommences, the latter is not invariably the initial point of the sequence, as in the first case.

b. In the second of those cases the interruptor causes a break in the keying sequence and a recommencement at any one of the 10 keying elements. Consequently, it is impossible now merely to superimpose sections of the text by shifting them so that their initial letters fall in the same column. But a superimposition is nevertheless possible, provided the interruptions do not occur so frequently [2] that sections of only a very few letters are enciphered by sequent keyletters. In order to accomplish a proper superimposition in this case, a statistical test is essential, and for this a good many letters are required. The nature of this test will be explained in Section XI.

c. The same thing is true of the last of the three cases mentioned under paragraph 18. The solution of a case of this sort is admittedly a rather difficult matter which will be taken up in its proper place later.

d. (1) In the cases thus far studied, either the plain-text groupings were variable in length and were enciphered by a constant-length key, or the plain-text groupings were constant in

---

[2] When no interruptor or "influence letter" is used, the interruption or break in the keying sequence occurs after the encipherment of a definite number of letters. Once this number has been ascertained, solution of subsequent messages is very simple.

length and were enciphered by a variable-length key. It is possible, however, to combine both principles and to apply a variable-length key to variable-length groupings of the plain text.

(2) Suppose the correspondents agree to encipher a message according to word lengths, but at irregular intervals, to add at the end of a word an interruptor letter which will serve to interrupt the key. Note the following, in which the key is BUSINESS MACHINES and the interruptor letter is X:

```
Key.........           B              U           S               B
Plain.......   A M M U N I T I O N   F O R   F I R S T X   A R T I L L E R Y etc.
Cipher......   B T T R V O D O W V   E Q V   Z D F G J O   B H D O S S J H I
```

CRYPTOGRAM

BTTRV ODOWV EQVZD FGJOB HDOSS JHI . . . etc.

(3) The foregoing system is only a minor modification of the simple case of ordinary word length encipherment as explained in Section II. If standard cipher alphabets are used, the spasmodic interruption and the presence of the interruptor letter would cause no difficulty whatever, since the solution can be achieved mechanically, by completing the plain-component sequence. If mixed cipher alphabets are used, and the primary components are unknown, solution may be reached by following the procedure outlined in Sections II and III, with such modifications as are suitable to the case.

e. It is hardly necessary to point out that the foregoing types of aperiodic substitution are rather unsuitable for practical military usage. Encipherment is slow and subject to error. In some cases encipherment can be accomplished only by single-letter operation. For if the interruptor is a cipher letter the key is interrupted by a letter which cannot be known in advance; if the interruptor is a plain-text letter, while the interruptions can be indicated before encipherment is begun, the irregularities occasioned by the interruptions in keying cause confusion and quite materially retard the enciphering process. In deciphering, the rate of speed would be just as slow in either method. It is obvious that one of the principal disadvantages in all these methods is that if an error in transmission is made, if some letters are omitted, or if anything happens to the interruptor letter, the message becomes difficult or impossible to decryptograph by the ordinary code clerk. Finally, the degree of cryptographic security attainable by most of these methods is not sufficient for military purposes.

SECTION VI

REVIEW OF AUTO-KEY SYSTEMS

23. The two basic methods of auto-key encipherment.—*a.* In auto-key encipherment there are two possible sources for successive key letters: the plain text or the cipher text of the message itself. In either case, the *initial* key letter or key letters are supplied by preagreement between the correspondents; after that the text letters that are to serve as the key are displaced 1, 2, 3, . . . intervals to the right, depending upon the length of the prearranged key.

b. (1) An example of plain-text keying will first be shown, to refresh the student's recollection. Let the previously agreed upon key consist of a single letter, say X, and let the cipher alphabets be direct standard alphabets.

```
Key..............  X N O T I F Y Q U A R T E R M A S T E R . .
Plain............  N O T I F Y Q U A R T E R M A S T E R . . .
Cipher...........  K B H B N D O K U R K X V D M S L X V . . .
```

(2) Instead of having a single letter serve as the initial key, a word or even a long phrase may be used. Thus (using TYPEWRITER as the initial key):

```
Key..............  T Y P E W R I T E R|N O T I F Y Q U A R . .
Plain............  N O T I F Y Q U A R T E R M A S T E R . . .
Cipher...........  G M I M B P Y N E I G S K U F Q J Y R . . .
```

c. (1) In cipher text auto keying the procedure is quite similar. If a single initial key letter is used:

```
Key..............  X|K Y R Z E C S M M D W A R D D V O S . . .
Plain............  N O T I F Y Q U A R T E R M A S T E R . . .
Cipher...........  K Y R Z E C S M M D W A R D D V O S J . . .
```

(2) If a key word is used:

```
Key..............  T Y P E W R I T E R|G M I M B P Y N E I . .
Plain............  N O T I F Y Q U A R T E R M A S T E R . . .
Cipher...........  G M I M B P Y N E I Z Q Z Y B H R R V . . .
```

(3) Sometimes only the last cipher letter resulting from the use of the prearranged key word is used as the key letter for enciphering the auto-keyed portion of the text. Thus, in the last example, the plain text beginning TERMASTER would be enciphered as follows:

```
Key..............  T Y P E W R I T E R|I B F W I I A T X . . .
Plain............  N O T I F Y Q U A R T E R M A S T E R . . .
Cipher...........  G M I M B P Y N E I B F W I I A T X O . . .
```

d. In the foregoing examples, direct standard alphabets are employed; but mixed alphabets, either interrelated or independent, may be used just as readily. Also, instead of the ordinary type of cipher alphabets, one may employ a mathematical process of addition (see par. 40*f* of Special Text No. 166, *Advanced Military Cryptography*) but the difference between the latter process and the ordinary one using sliding alphabets is more apparent than real.

e. Since the analysis of the case in which the cipher text constitutes the auto key is usually easier than that in which the plain text serves this function, the former will be the first to be discussed.

462861 O - 42 - 3

Section VII

SOLUTION OF CIPHER-TEXT AUTO-KEY SYSTEMS

24. Solution of cipher-text auto-keyed cryptograms when known alphabets are employed.— *a.* (1) First of all it is to be noted that if the cryptanalyst knows the cipher alphabets which were employed in encipherment, the solution presents hardly any problem at all. It is only necessary to decipher the message beyond the key letter or key-word portion and the initial part of the plain text enciphered by this key letter or key word can be filled in from the context. An example, using standard cipher alphabets, follows herewith:

Cryptogram

WSGQV OHVMQ WEQUH AALNB NZZMP ESKD

(2) Writing the cipher text as key letters (displaced one interval to the right) and deciphering by direct standard alphabets yields the following:

```
Key........     W S G Q V O H V M Q W E Q U H A A L N B N Z Z M P E S K
Cipher.....   W S G Q V O H V M Q W E Q U H A A L N B N Z Z M P E S K D
Plain......     W O K F T T O R E G I M E N T A L C O M M A N D P O S T
```

(3) Trial of the word REPORT as the initial word of the message yields an intelligible word as the initial key: FORCE, so that the message reads:

```
Key........   F O R C E V O H V M Q . .
Cipher.....   W S G Q V O H V M Q . . .
Plain......   R E P O R T T O R E . . .
```

(4) A semiautomatic method of solving such a message is to use sliding normal alphabets and align the strips so that, as one progresses from left to right, each cipher letter is set opposite the letter A on the preceding strip. Taking the letters VMQWEQUHA in the foregoing example, note in Figure 7 the series of placements of the successive strips. Then note how the successive plain-text letters of the word REGIMENT reappear to the left of the successive cipher letters MQWEQUHA.

A	V	H	X	T	X	N	H	O
B	W	I	Y	U	Y	O	I	P
C	X	J	Z	V	Z	P	J	Q
D	Y	K	A	W	A	Q	K	R
E	Z	L	B	X	B	R	L	S
F	A	M	C	Y	C	S	M	T
G	B	N	D	Z	D	T	N	U
H	C	O	E	A	E	U	O	V
I	D	P	F	B	F	V	P	W
J	E	Q	G	C	G	W	Q	X
K	F	R	H	D	H	X	R	Y
L	G	S	I	E	I	Y	S	Z
M	H	T	J	F	J	Z	T	A
N	I	U	K	G	K	A	U	B
O	J	V	L	H	L	B	V	C
P	K	W	M	I	M	C	W	D
Q	L	X	N	J	N	D	X	E
R	M	Y	O	K	O	E	Y	F
S	N	Z	P	L	P	F	Z	E
T	O	A	Q	M	Q	G	A	H
U	P	B	R	N	R	H	B	I
V	Q	C	S	O	S	I	C	J
W	R	D	T	P	T	J	D	K
X	S	E	U	Q	U	K	E	L
Y	T	F	V	R	V	L	F	M
Z	U	G	W	S	W	M	G	N

FIGURE 7.

b. If, as a result of the analysis of several messages (as described in par. 25), mixed primary components have been reconstructed, the solution of subsequent messages may readily be accomplished by following the procedure outlined in *a* above, since in that case the cipher alphabets have become known alphabets.

25. General principles underlying solution of cipher-text auto-keyed cryptograms by frequency analysis.—*a.* First of all, it is to be noted in connection with cipher-text auto-keying that repetitions will not be nearly as plentiful in the cipher text as they are in the plain text, because in this system before a repetition can appear two things must happen simultaneously. First, of course, the plain-text sequence must be repeated, and second, one or more cipher-text letters (depending upon the length of the introductory key) immediately before the second appearance of the plain-text repetition must be identical with one or more cipher-text letters immediately before the first appearance of the group. This can happen only as the result of chance. In the following example the introductory key is a single letter, X, and direct standard components are used in the usual Vigenère manner:

Key	X C K B T M D H N V H L Y . . . . K D K S J M D H N V H L Y
Plain	F I R S T R E G I M E N T . . . . T H I R D R E G I M E N T
Cipher	C K B T M D H N V H L Y R . . . K D K S J M D H N V H L Y R

The repeated plain-text word, REGIMENT, has only 8 letters but the repeated cipher-text group contains 9, of which only the last 8 letters actually represent the plain-text repetition. In order that the word REGIMENT be enciphered by D H N V H L Y R the second time this word appeared in the text it was necessary that the key letter for its first letter, R, be M *both* times; no other key letter will produce the same cipher sequence for the word REGIMENT in this case. Each different key letter for enciphering the first letter of REGIMENT will produce a different encipherment for the word, so that the chances [1] for a repetition in this case are roughly about 1 in 26. This is the principal cause for the reduction in repetitions in this system. If an introductory key of two letters were used, it would be necessary that the two cipher letters immediately before the second appearance of the repeated word REGIMENT be identical with the two cipher letters immediately before the first appearance of the word. In general, then, an n-letter repetition in the cipher text, in this case, represents an $(n-k)$-letter repetition in the plain text, where n is the length of the cipher-text repetition and k is the length of the introductory key.

b. There is a second phenomenon of interest in connection with the cipher-text auto-key method. Let the letter opposite which the key letter is placed (when using sliding components for encipherment) be termed, for convenience in reference, "the base letter." Normally the base letter is the initial letter of the plain component, but it has been seen in preceding texts that this is only a convention. Now when the introductory key is a single letter, if the base letter occurs as a plain-text letter its cipher equivalent is identical with the immediately preceding cipher letter; that is, there is produced a double letter in the cipher text, no matter what the cipher component is and no matter what the key letter happens to be for that encipherment. For example, using the H Y D R A U L I C . . . X Z sequence for both primary components, with H, the initial letter of the plain component as the base letter, and using the introductory key letter X, the following encipherment is produced:

Key	X	J	O	I	I	F	L	Y	U	T	T	D	K	K	Y	C	X	G
Plain	M	A	N	H	A	T	T	A	N	H	I	G	H	J	I	N	K	S
Cipher	J	O	<u>I</u>	<u>I</u>	F	L	Y	U	<u>T</u>	<u>T</u>	D	<u>K</u>	<u>K</u>	Y	C	X	G	L

Note the doublets II, TT, KK. Each time such a doublet occurs it means that the second letter represents H_p, which is the base letter in this case (initial letter of plain component). Now if the base letter happens to be a high-frequency letter in normal plain text, for example the letter E, or T, then the cipher text will show a large number of doublets; if it happens to be a low-frequency letter the cipher text will show very few doublets. In fact, the number of doublets will be directly proportional to the frequency of the base letter in normal plain text. Thus, if the cryptogram contains 1,000 letters there should be about 72 occurrences of doublets if the base letter is A, since in 1,000 letters of plain text there should be about 72 A's. Conversely, if a cryptogram of 1,000 letters shows about 72 doublets, the base letter is likely to be A; if it shows about 90, it is likely to be T, and so on. Furthermore when a clue to the identity of the base letter has been obtained in this manner, it is possible immediately to insert the corresponding plain-text letter throughout the text of the message. The distribution of this letter may not only serve as a check (if no inconsistencies develop) but also may lead to the assumption of values for other cipher letters.

c. When the introductory key is 2 letters, then this same phenomenon will produce groups of the formula ABA, where A and B may be any letters, but the first and third must be identical. The occurrence of patterns of this type in this case indicates the encipherment of the base letter.

[1] If all the cipher letters appeared with equal frequency the chances would be exactly 1 in 26. But certain letters appear with greater frequency because some plain-text letters are much more frequent than others.

d. The phenomena noted above can be used to considerable advantage in the solution of cryptograms of this type. For instance, if it is known that the ordinary Vigenère method of encipherment is used ($\Theta_{k/2}=\Theta_{1/1}$; $\Theta_{p/1}=\Theta_{c/2}$), then the initial letter of the plain component is the base letter. If, further, it is known that the plain component is the normal direct sequence, then the base letter is A and a word such as BATTALION will be enciphered by a group having the formula AABCCDEFG. If the plain component is a mixed sequence and happens to start with the letter E, then a word such as ENEMY would be enciphered by a sequence having the formula AABBCD.[2] Sequences such as these are, of course, idiomorphic and if words yielding such idiomorphisms are frequent in the text there will be produced in the latter several or many cases of isomorphism. When these are analyzed by the principles of indirect symmetry of position, a quick solution may follow.

e. A final principle underlying the solution of cipher-text auto-keyed cryptograms remains to be discussed. It concerns the nature of the frequency distributions required for the analysis of such cryptograms. This principle will be set forth in the next paragraph.

26. Frequency distributions required for solution.—*a*. Consider the message given in paragraph 23*c* (1). It happens that the letter R_c occurs twice in this short message and, because of the nature of the cipher-text auto-keying method, this letter must also appear twice in the key. Now it is obvious that all plain-text letters enciphered by key letter R_k will be in the same cipher alphabet; in other words, if the key text is "offset" one letter to the right of the cipher text, *then every cipher letter which immediately follows an* R_c *in the cryptogram will belong to the same cipher alphabet*, and this alphabet may be designated conveniently as the R cipher alphabet. Now if there were sufficient text, so that there were, say, 30 to 40 R_c's in it, then a frequency distribution of the letters immediately following the R_c's will exhibit monoalphabeticity. What has been said of the letters following the R_c's applies equally well to the letters following all the other letters of the cipher text, the A_c's, B_c's, C_c's, and so on. In short, if 26 distributions are made, one for each letter of the alphabet, showing the cipher letter immediately succeeding each different letter of the cipher text, then the text of the cryptogram can be allocated into 26 uniliteral, monoalphabetic frequency distributions which can be solved by frequency analysis, providing there are sufficient data for this purpose.

b. The foregoing principle has been described as pertaining to the case when the introductory key is a single letter, that is, when the key text is "offset" or displaced but one interval to the right of the cipher text. But it applies equally to cases wherein the key text is offset more than one interval, provided the frequency distributions are based upon the proper interval, as determined by the displacement due to the length of the introductory key. For instance, suppose the introductory key consists of two letters, as in the following example:

```
Key text..........  X Z|M R H F H G F N Q R X O M R M V W E E
Plain text........  R E L I A B L E I N F O R M A T I O N . .
Cipher text.......  M R H F H G F N Q R X O M R M V W E E . .
```

The key text in this case is offset two intervals to the right of the cipher text and, therefore, frequency distributions made by taking the cipher letters one interval to the right of a given cipher letter, each time that letter occurs, will not be monoalphabetic because some letter not related at all to the given cipher letter is the key letter for enciphering the letter one interval to the right of the latter. For example, note the three R_c's in the foregoing illustration. The first R_c is followed by H_c, representing the encipherment of L_p by M_k; the second R_c is followed by X_c, representing the encipherment of F_p by Q_k; the third R_c is followed by M_c, representing the encipherment of A_p by M_k. The three cipher letters H, X, and M are here entirely unrelated and do

[2] Six letters are shown because the idiomorphism in this case extends over that many letters.

not belong to the same cipher alphabet because they represent encipherments by three different key letters. On the other hand, the cipher letters two intervals to the right of the R_e's, viz, F, O, and V, are in the same cipher alphabet because these cipher letters are the results of enciphering plain-text letters I, O, and T, respectively, by the *same* key letter, R. It is obvious, then, that when the introductory key consists of two letters and the key text is displaced two intervals to the right of the cipher text, the proper frequency distributions for monoalphabeticity will be based upon the letter at the second interval to the right of each cipher letter. Likewise, if the introductory key consists of three letters and the key text is displaced three intervals to the right of the cipher text, the distributions must be based upon the third interval, and so on, in each case the interval used corresponding to the amount of displacement between key text and cipher text.

c. Conversely, in solving a problem of this type, when the length of the introductory key and therefore the amount of displacement are not known, the appearance of the frequency distributions based upon various intervals after each different cipher letter will disclose this unknown factor, since only one set of distributions will exhibit monoalphabeticity and the interval corresponding to that set will be the correct interval.

d. Application of these principles will now be made, using a specific example.

27. Example of solution by frequency analysis.—*a.* It will be assumed that previous studies have disclosed that the enemy is using the cipher-text auto-key system described. It will be further assumed that these studies have also disclosed that (1) the introductory key is usually a single letter, (2) the usual Vigenère method of employing sliding primary components is used, (3) the plain component is usually the normal direct sequence, the cipher component a mixed sequence which changes daily. The following cryptograms, all of the same date, have been intercepted:

MESSAGE I

```
I J X W X   E E C D A   C N Q E T   U K N M V   D I W P P
Q Z S X D   H I F E L   N N J J I   D I V E Y   G T C Z M
E H H L M   R V C U R   G D I E Q   S G T A R   J J Q Q Y
C A R P H   M G L D Y   F Y T C D   G Y F K R   F K S E T
T D I Q K   K M L T U   R Q G G N   K M K I X   J X W K A
O K N T B   T Z J O Q   Y S C D I   D G E T X   G X X X X
```

MESSAGE II

```
G R V R M   Z W K X G   W P C K K   R M X A N   J C C X U
R T N J U   A K O B L   N L M W K   Y Y Z J U   C S U H F
F H I J A   Q B M L T   P U R R S   U E Q E V   Z E Y G C
F F N F I   B W N Y S   T C E T P   D G T T Z   R R Q H Q
A O O X D   B U Y N K   L B W C D   G G K X X
```

MESSAGE III

```
R W K A O   L T C J M   Z D K V U   J C D D Y   B Z E L M
M W T Q O   H Q V G X   C H O L M   W V G R K   I B R X D
L A Q Y U   K I R O Z   T Q Y U X
```

Message IV

```
X J J P M   L T Z K X   E C A Q Z   N T T O C   O N D U C
T U T C V   G R J P F   F D I P P   D I X C E   S E T W W
S U M U J   C S L G X   H X M O Z   E K A Q I   S U A O X
```

Message V

```
G I S U H   W Z H S T   T Z O I D   D H O O V   N B T J G
X C T B S   F K I R H   M M V Y M   I I V U U   C Z M J E
H A G I E   W M E H H   L M W K Y   P P D Q Z   G B O I W
P S F A J   U Q Z H Z   M T F H Z   M L A C Z   R O V D I
W P V I B   O B C C X   N N D G I   E S J O C   K B J H Q
M U Z E L   Y O O V U   J W K I E   I B B C Z   A J I E F
F O R S A   J L N Q M   B Q X X X
```

Message VI

```
T B J P A   A R Y Y P   V H I D I   T U X N J   M X G S S
B D A Q Y   M M T T F   U U N M G   Q P U X M   O V U Y E
C E C Z M   M W O H C   F O B H V   N K A Z C   K M X X X
```

Message VII

```
T B J P A   Q A A Z T   R X A L X   F K K M E   I A A B D
S F T Q T   C J J G J   O V M R G   L V W T T   J U A W L
X U K T X   G G B O X   M X D I D   S P B S F   L Y Z K C
F X X X X
```

b. A distribution table of the type described in paragraph 25*e* is compiled and is shown as Figure 8 below. In making these distributions it is simple to insert a tally in the appropriate cell in the pertinent horizontal line of the table, to indicate the cipher letter which immediately follows each occurrence of the letter to which that line applies. Obviously, the best method of compiling the data is to handle the text digraphically, taking the first and second letters, the second and third, the third and fourth, and so on, and distributing the final letters of the digraphs in a quadricular table. The distribution merely takes the form of tally marks, the fifth being a diagonal stroke so as to totalize the occurrences visibly.

SECOND LETTER

FIRST LETTER	A	B	C	D	E	F	G	H	I	J	K	L	M	N	O	P	Q	R	S	T	U	V	W	X	Y	Z
A	///	/	//				/			///	/	/		/	////		///// /	///					/			//
B		/	/	///				/		///		/	/		////		/	/	//	///	/		//			/
C	//		//	/////	///	///		/		//	///			/					//	/	/	/		///		////
D	//	/		//		/	/////	//	///// /////		/	/					/		//						//	
E			////		/	/		///	//		/	///					//		//	/////		/	/		//	
F	/			/	/	////		//	/	/	////	/		/	//					/	/				/	
G		//	/	/	/		///		///	/	/	//		/			/	///	/	///			/	//	/	
H	/		/			/		//	///			//	/		//		///		/			/	/	/		//
I	/	////		/////	/////	/			/	//						/	/	//	//	/		//	///	//		
J	/		////		/		/	/	//	////		/	///		///	////	/				///		/	//		
K	////	/	/						/////		///	/	////	//	/			//	/	/		/		//	//	
L	//	/		/			/						/////	///						////		/		//	//	
M		///			///		///		/	/	/	////	////		//			//		//	//	//	///// /	///		//
N		/		/		/				////	///	/	//	//			//			/					/	
O		///	/					//	//		/	//			///		/	/				/////		//		///
P	//	/	/	///		/		/					/			///	/		/		//	//				
Q	//	/			//		/	/	/		/		//		/	/	/		/	/		/			/////	////
R						/	//	/		//	/		//		//	/	//	//	//	/		//	/	//	/	
S	/	/	/		//	////	/			/		/				/			/	//	/////			/		
T	/	////	///// /	/		//				//				/		//	///	/		/////	////		/	//		////
U	///		///		/			//		///	///		/	/			/	////		/	//			//	//	/
V			/	//	/		///	/	/				/	//				/			////		/		/	/
W			/								///// /	/	/	/	/	////			/	//		/	/	//		/
X	//		///	////	//	/	////	/		//			////	//							//		//			
Y		/	/		/	//	//						//	/	/	//			//	/	//				//	//
Z	/		/	/	////		/	//		//	//		///// /		/			//	/	//			/			

FIGURE 8.

c. The individual frequency distributions give every appearance of being monoalphabetic, which checks the assumption that the enemy is still employing the same system. The total number of letters of text (excluding the final X's) is 680. If the base letter is A then there should be approximately 680×7.2%=49 cases of double letters in the text. There are actually 52 such cases, which checks quite well with expectancy. The letter A is substituted throughout the text for the second letter of each doublet.

d. The following sequence is noted:

```
Message V, line 1........ G I S U H  W Z H S T  T Z O I D  D H O O V  N B T J G
                                           . . . A . . . . A . . A . . .
```

Assume that the sequence DDHOOVNBT represents BATTALION. Then the frequency of H_c in the D cipher alphabet should be high, since $H_c=T_p$. The H has only 2 occurrences. Likewise, the frequency of O_c in the H alphabet ($=T_p$) should be high; it is also only 2. The frequency of V in the O alphabet should be medium or low, since it would equal L_p; it is 5, which is too high. The rest of the letters of the assumed word are similarly checked against the appropriate frequency distributions, with the result that, on the whole, the assumption that the DDHOOVNBT sequence represents BATTALION does not appear to be warranted. Similar attempts are made at other points in the text, with the same or other probable words. Some of these attempts may have to be carried to the point where the placement of values in the tentative cipher component leads to serious inconsistencies. Finally, attention is fixed upon the following sequence:

Message VI, line 2........ B D A Q Y M M T T F U U N M G . . .
. . . A . A . . A . . .

The word MMTTFUUNMG / AVAILABLE is assumed. The appropriate frequency distributions are consulted to see how well the actual individual frequencies correspond to the expected ones.

Alphabet	Assumed θ_c	Assumed θ_p	Frequency Expected	Frequency Actual	Approximation
M	T	V	Low	2	Fair
T	F	I	High	2	Fair
F	U	L	Medium	1	Good
U	N	B	Low	1	Good
N	M	L	Medium	2	Fair
M	G	E	High	3	Fair

The assumption cannot be discarded just yet. Let the values derivable from the assumption be inserted in their proper places in a cipher component, and, using the latter in conjunction with a normal direct sequence as the plain component, let an attempt be made to find corroboration for these values. The following placements may be made:

Plain................. A B C D E F G H I J K L M N O P Q R S T U V W X Y Z
Cipher................ M F G U N T

The letter M_c appears twice in the cipher sequence and when this partially reconstructed cipher component is tested it is found that the value $L_p(N_k)=M_c$ is corroborated. Having the letters M, F, G, U, N, and T tentatively placed in the cipher component, it is possible to insert certain plain-text values in the text. For example, in the M alphabet, $F_c=D_p$, $G_c=E_p$, $U_c=O_p$, $N_c=P_p$, $T_c=V_p$. In the F alphabet, $G_c=B_p$, $U_c=L_p$, $N_c=M_p$, $T_c=S_p$, $M_c=X_p$. The other letters yield additional values in the appropriate alphabets. The plain-text values thus obtainable are inserted in the cipher text. No inconsistencies appear and, moreover, certain "good" digraphs are brought to light. For instance, note what happens here:

Message V, line 4..................
Key............. . U Q Z H Z M T F H Z M L A C Z
Cipher.......... U Q Z H Z M T F H Z M L A C Z .
Plain........... V I

Now if the letter H can be placed in the cipher component, several values might be added to this partial decipherment. Noting that F and G are sequent in the cipher component, suppose H follows G therein. Then the following is obtained:

Message V, line 4..................
Key............. . U Q Z H Z M T F H Z M L A C Z
Cipher.......... U Q Z H Z M T F H Z M L A C Z .
Plain........... V I C

Suppose the VIC is the beginning of VICINITY. This assumption permits the placement of A, C, L, and Z in the cipher component, as follows:

	A	B	C	D	E	F	G	H	I	J	K	L	M	N	O	P	Q	R	S	T	U	V	W	X	Y	Z
Plain	A	B	C	D	E	F	G	H	I	J	K	L	M	N	O	P	Q	R	S	T	U	V	W	X	Y	Z
Cipher	M	A		F	G	H			L					Z	U	N						T				C

These additional values check in very nicely and presently the entire cipher component is reconstructed. It is found to be as follows:

Plain	A	B	C	D	E	F	G	H	I	J	K	L	M	N	O	P	Q	R	S	T	U	V	W	X	Y	Z
Cipher	M	A	B	F	G	H	J	K	L	Q	S	V	X	Z	U	N	D	E	R	W	O	T	Y	P	I	C

The key phrase is obviously UNDERWOOD TYPEWRITER COMPANY. All the messages now may be deciphered with ease. The following gives the letter-for-letter decipherment of the first three groups of each message:

I (Introductory key: K)

Key	K I J X W	X E E C D	A C N C Q	...
Cipher	T J X W X	E E C D A	C N Q E T	...
Plain	R I G H T	F A I R L	Y Q U I E	...

II (Introductory key: E)

Key	E G R V R	M Z W K X	G W P C K	...
Cipher	G R V R M	Z W K X G	W P C K K	...
Plain	N O T H I	N G O F S	P E C I A	...

III (Introductory key: R)

Key	R R W K A	O L T C J	M Z D K V	...
Cipher	R W K A O	L T C J M	Z D K V U	...
Plain	A B O U T	O N E H U	N D R E D	...

IV (Introductory key: J)

Key	J X J J P	M L T Z K	X E C A Q	...
Cipher	X J J P M	L T Z K X	E C A Q Z	...
Plain	G U A R D	I N S U F	F I C I E	...

V (Introductory key: E)

Key	E G I S U	H W Z H S	T T Z O I	...
Cipher	G I S U H	W Z H S T	T Z O I D	...
Plain	N U M E R	O U S F L	A S H E S	...

VI (Introductory key: B)

Key	B T B J P	A A R Y Y	P V H I D	...
Cipher	T B J P A	A R Y Y P	V H I D I	...
Plain	T H E R E	A R E A B	O U T S I	...

VII (Introductory key: B)

Key	B T B J P	A Q A A Z	T R X A L	...
Cipher	T B J P A	Q A A Z T	R X A L X	...
Plain	T H E R E	I S A M I	X U P H E	...

e. In the foregoing example the plain component was the normal direct sequence, so that with the Vigenère method of encipherment the base letter is A. If the plain component is a mixed sequence, the base letter may no longer be A, but in accordance with the principle set forth in paragraph 25*b*, the frequency of doublets in the cipher text will correspond with the frequency of the base letter as a letter of normal plain text. If a good clue as to the identity of this letter is afforded by the frequency of doublets in the cipher text, the insertion of the corresponding base letter in the plain text will lead to further clues. The solution from there on can be handled along the lines indicated above.

28. Example of solution by analysis of isomorphisms.—*a*. It was stated in paragraph 25*d* that in cipher-text auto-keying the production of isomorphs is a frequent phenomenon and that analysis of these isomorphs may yield a quick solution. An example of this sort will now be studied.

b. Suppose the following cryptograms have been intercepted:

1

```
U S Y P W   T R X D I   M L E X R   K V D B D   D Q G S U   N S F B O
B E K V B   M A M M O   T X X B W   E N A X M   Q L Z I X   D I X G Z
P M Y U C   N E V V J   L K Z E K   U R C N I   F Q F N N   Y G S I J
T C V N I   X D D Q Q   E K K L R   V R F R F   X R O C S   S J T B V
E F A A G   Z R L F D   N D S C D   M P B B V   D E W R R   N Q I C H
A T N N B   O U P I T   J L X T C   V A O V E   Y J J L K   D M L E G
N X Q W H   U V E V Y   P L Q G W   U P V K U   B M M L B   O A E O T
T N K K U   X L O D L   W T H C Z   R
```

2

```
B I I B F   G R X L G   H O U Z O   L L Z N A   M H C T Y   S C A A T
X R S C T   K V B W K   O T G U Q   Q F J O C   Y Y B V K   I X D M T
K T T C F   K V K R O   B O E P L   Q I G N R   I Q O V J   Y K I P H
J O E Y M   R P E E W   H O T J O   C R I I X   O Z E T Z   N K
```

3

```
H A L O Z   J R R V M   M H C V B   Y U H A O   E O V A C   Q V V J L
K Z E K U   R F R F X   Y B H A L   Z O F H M   R S J Y L   A P G R S
X A G X D   M C U N X   X L X G Z   J P W U I   F D B B Y   P V F Z N
B J N N B   I T M L J   O O S E A   A T K P B   Y
```

c. Frequency distributions are made, based upon the 2d letters of pairs, as in the preceding example. The result is shown in the table in figure 9. The data in each distribution are relatively scanty and it would appear that the solution is going to be a rather difficult matter.

SECOND LETTER

FIRST LETTER

	A	B	C	D	E	F	G	H	I	J	K	L	M	N	O	P	Q	R	S	T	U	V	W	X	Y	Z	
A	///		/		/		//					//	//		//	/				///				/			A
B		//		/	/	/		/	//	/			//		////							///	//		///		B
C	/			/		/		/						//			/	/	/	//	/	///			/	/	C
D		//		//	/				//			/	////	/			//		/								D
E	/				/	/	/				////			/	//	/				/		//	//	/	//		E
F	/	/		//			/	/		/	/			/			/	//						//		/	F
G								/						//				//	//		/		/	/		///	G
H	////		///							/			/		//						/						H
I			/			//	/	/	/				/			/	/			//				~~////~~			I
J										/		////		/	////	/		/		//					//		J
K				/					//		//	/			/	/		/		/	////	////				//	K
L	/	/			//	/	/			/	///	/			///		//	/					/	//		///	L
M	/		/					//				////	///		/	/	/	//		/					/		M
N	//	///		/	/				//		//			///			/	/	/					//	/		N
O	/	//	///	/	///	/						/			/				/	////	//	///				//	O
P		//			/		/	/	/			//	/									//	//				P
Q					/	//	//		//			/			/		//					/	/				Q
R			/			////			//		/	/		/	//	/		//	///			//		//			R
S			///		/	/			/	//									/		/			/	/		S
T		/	///				/	/		//	///		/	//				/		//				//	/	/	T
U		/	/					/	/					//		//	/	//	/			/		/		/	U
V	//	///		//	///	/				///	///		/	/				/				//			/		V
W		/			/			//			/							/		//	//						W
X	/	/		~~////~~			//					///	/		/		/	///		/				//	/		X
Y		//					/			/	/	/	/			///			/		//				/		Y
Z					///				/	//				///	//	/		//									Z
	A	B	C	D	E	F	G	H	I	J	K	L	M	N	O	P	Q	R	S	T	U	V	W	X	Y	Z	

FIGURE 9.

d. However, before becoming discouraged too quickly, a search is made throughout the text to see if any isomorphs are present. Fortunately there appear to be several of them. Note the following:

Message 1	(1)	. D B D D Q G S U N S F B O B E K . . .
	(2)	. N E V V J L K Z E K U R C N I F . . .
	(3)	. T N K K U X L O D L W T H C Z R | end of message
Message 2	(4)	. C R I I X O Z E T Z N K | end of message
Message 3	(5)	. C Q V V J L K Z E K U R F R F X . . .

First, it is necessary to delimit the length of the isomorphs. Isomorph (2) shows that the isomorphism begins with the doubled letters. For there is an E before the V V in that case and also an E within the isomorph; if the phenomenon included the E, then the letter immediately before the D D in the case of isomorph (1) would have to be an N, to match its homolog, E, in isomorph (2), which it is not. Corroborating data are given by isomorphs (3), (4), and (5) in this respect. Hence, we may take it as established that the isomorphism begins with the doubled letters.

As for the end of the isomorphism, the fact that isomorphs (2) and (5) are the same for 10 letters seems to indicate that that is the length of the isomorphism. The fact that message 2 ends 2 letters after the last "tie-in" letter, Z, corroborates this assumption. It is at least certain that the isomorphism does not extend beyond 11 letters because the recurrence of R in isomorph (5) is not matched by the recurrence of R in isomorph (2), nor by the recurrence of T in isomorph (3). Hence it may be assumed that the isomorphic sequence is probably 10 letters in length, possibly 11. But to be on safe ground it is best to proceed on the 10-letter basis.

e. Applying the principles of indirect symmetry to the superimposed isomorphs, partial chains of equivalents may be constructed and it happens in this case that practically the entire primary component may be established. Let the student confirm the fact that the following sequence may be derived from the data given:

1	2	3	4	5	6	7	8	9	10	11	12	13	14	15	16	17	18	19	20	21	22	23	24	25	26
T	E	Z	K	R	.	I	V	F	.	.	.	Q	.	W	G	.	N	U	S	B	X	J	D	O	L

The only missing letters are A, C, H, M, P, and Y. By use of the nearly complete sequence on the text it will be possible to place these 6 letters in their positions in the cipher component. Or, if a keyword-mixed sequence is suspected, then the sequence which was reconstructed may be merely a decimation of the original primary sequence. By testing the partial sequence for various intervals, when the seventh is selected the following result is obtained:

1	2	3	4	5	6	7	8	9	10	11	12	13	14	15	16	17	18	19	20	21	22	23	24	25	26
T	V	W	X	Z	.	.	D	R	.	U	L	I	.	B	E	F	G	J	K	.	N	O	.	Q	S

The sequence is obviously based on the keyword HYDRAULIC, and the complete primary cipher component is now available. The plain component is then to be reconstructed. A word must be assumed in the text.

f. A good probable word to assume for the 10-letter repetition found in messages 1 and 3 is ARTILLERY. This single assumption is sufficient to place 7 letters in the plain component. Thus:

```
Key............... . . . V V J L K Z E K U R . . .
Plain............. . . . A R T I L L E R Y . . . .
Cipher............ . . V V J L K Z E K U R . . . .
```

1	2	3	4	5	6	7	8	9	10	11	12	13	14	15	16	17	18	19	20	21	22	23	24	25	26
A	.	.	.	E	.	.	.	I	.	.	L	.	.	.	.	.	R	.	T	.	.	.	.	Y	.

These few letters are sufficient to indicate that the plain component is probably the normal direct sequence. A few minutes testing proves this to be true. The two components are therefore:

```
Plain............. A B C D E F G H I J K L M N O P Q R S T U V W X Y Z
Cipher............ H Y D R A U L I C B E F G J K M N O P Q S T V W X Z
```

With these two components at hand, the decipherment of the messages now becomes a relatively simple matter. Assuming a single-letter introductory key, and trying the first five groups of message 1 the results are as follows:

```
Key............... ? U S Y P   W T R X D   I M L E X   R K V D B   D D Q G S   . . .
Cipher............ U S Y P W   T R X D I   M L E X R   K V D B D   D Q G S U   . . .
Plain............. ? P H R F   Y I V E F   I R E O F   L I G H T   A R T I L   . . .
```

It is obvious that an introductory key of more than one letter was used, since the first few letters yield unintelligible text; but it also appears that the last cipher letter of the introductory key was used as the introductory key letter for enciphering the subsequent auto-keyed portion of the text (see par. 23*c*(3)). However, assuming that the IVE before the word FIRE is the ending of the first word of the plain text, and that this word is INTENSIVE, the introductory key word is found to be WICKER. Thus:

```
Key.......... W I C K E R|T R X D I M L E X R K V D B D D Q G S . . .
Plain........ I N T E N S I V E F I R E O F L I G H T A R T I L . . .
Cipher....... U S Y P W T R X D I M L E X R K V D B D D Q G S U . . .
```

The beginnings of the other two messages are recoverable in the same way and are found to be as follows:

```
Key................ P R O M I S E|R X L G H O U Z O . .
Plain.............. R E Q U E S T V I G O R O U S . . .
Cipher............. B I I B F G R X L G H O U Z O . . .

Key................ C H A R G E D|R R V M M H C V B . . .
Plain.............. S E C O N D B A T T A L I O N . . . .
Cipher............. H A L O Z J R R V M M H C V B . . . .
```

g. The example solved in the foregoing subparagraphs offers an important lesson to the student, insofar as it teaches him that *he should not immediately feel discouraged when confronted with a problem presenting only a small quantity of text and therefore affording what seems at first glance to be an insufficient quantity of data for solution.* For in this example, while it is true that there are insufficient data for analysis by simple principles of frequency, it turned out that solution was achieved *without any recourse to the principles of frequency of occurrence.* Here, then, is one of those interesting cases of substitution ciphers of rather complex construction which are solvable without any study whatsoever of frequency distributions. Indeed, it will be found to be true that in more than a few instances the solution of quite complicated cipher systems may be accomplished not by the application of the principles of frequency, but by recourse to inductive and deductive reasoning based upon other considerations, even though the latter may often appear to be very tenuous and to rest upon quite flimsy supports.

29. Special case of solution of cipher-text auto-keyed cryptograms.—*a*. Two messages with identical plain texts enciphered according to the method of paragraph 23 *c* (3) by initial key words of different lengths and compositions can be solved very rapidly by reconstructing the primary components. The *cryptographic texts of such messages will be isomorphic after the initial key-word portions.* Note the two following superimposed messages, in which isomorphism between the two cryptograms is obvious after their 6th letters:

```
1. T S B J S   K B N L O   C F H A Z   L W J A M   B N F N S   M V J R E
2. B K K M J   X Y C X B   H R P V O   X M U V I   Y C R C G   I K U T D

1. H F P R X   C P C R R   E H F M U   H R A X C   N F D U B   A T F Q R
2. P R E T N   H E H T T   D P R I W   P T V N H   C R S W Y   V J R F T
```

Starting with any pair of superimposed letters (beginning with the 7th pair), chains of equivalents are constructed:

	1	2	3	4	5	6	7	8	9	10	11	12	13	14
1	Z	O	B	Y	.	.	.							
2	L	X	N	C	H	P	E	D	S	G	.	.	.	.
3	Q	F	R	T	J	U	W	M	I	.	.	.		
4	A	V	K	.	.	.								

By interpolation, these partial sequences may be united into the key-word sequence:

H Y D R A U L I C B E F G J K M N O P Q S T V W X Z

b. The initial key words and the plain texts may now be ascertained quite easily by deciphering the messages, using this primary component slid against itself. It will be found that the initial key word for the 1st message is PENCE, that for the 2d is LATERAL. The reason that the cryptographic texts are isomorphic beyond the initial key word portions is, of course, that since the text beyond the key word is enciphered auto-key fashion by the preceding cipher letter the letters before the last letter of the key have no effect upon the encipherment at all. Hence two messages of identical text cannot be other than isomorphic after the initial key-word portions.

c. The foregoing solution affords a clue to the solution of cases in which the texts of two or more messages are not completely identical but are in part identical because they happen to have similar beginnings or endings, or contain nearly similar information or instructions. The progress in such cases is not so rapid as in the case of messages with wholly identical texts because much care must be exercised in blocking out the isomorphic sequences upon which the reconstruction of the primary components will be based.

d. (1) In the foregoing cases, the primary components used to encipher the illustrative messages were identical mixed sequences. If nonidentical components are employed, the cryptograms present an interesting case for the application of a principle pointed out in a preceding text.[4]

(2) Suppose that the three messages of paragraph 27*b* had been enciphered by using a plain component different from the mixed component. The encipherments of the word ARTILLERY would still yield isomorphic sequences, from which, as has been noted, the reconstruction of the cipher component can be accomplished.

(3) Having reconstructed the cipher component (or an equivalent) the latter may be applied to the cipher text and a "decipherment" obtained. In this process *any* sequence of 26 letters may be used as the plain component and even the normal sequence A . . . Z may be employed for this purpose. The word decipherment in the next to the last sentence is enclosed by quotation marks because the letters thus obtained would not yield plain text, since the real or an equivalent plain component has not yet been found. Such "deciphered" text may be termed *spurious* plain text. *But the important thing to note is that this text is now monoalphabetic and may be solved by the simple procedure usually employed in solving a monoalphabetic cipher produced by a single mixed alphabet.* Thus, a polyalphabetic cipher may be converted to monoalphabetic terms and the problem much simplified. In other words, here is another example of the situations in which the principle of conversion into monoalphabetic terms may be applied with gratifying success. It is also an example of the dictum that the use of two differently mixed primary components does not really give much more security than does a mixed component sliding against itself or against the normal sequence.

---

[4] *Military Cryptanalysis, Part II*, par. 45*g*.

e. (1) If the auto-key method shown in paragraph 23*c* (2) had been employed in enciphering the two identical texts above, the solution would, of course, have been a bit more difficult. To illustrate such a case, let the two texts be enciphered by key words of the same lengths but different compositions: PENCE and LATER. Thus:

No. 1

Key	P E N C E	T S B J S	M M N R U	L P U I H	J B T X F	I N N R M
Plain	R E Q U E	S T I N F	O R M A T	I O N O F	S I T U A	T I O N I
Cipher	T S B J S	M M N R U	L P U I H	J B T X F	I N N R M	D W I Q V
Key	D W I Q V	P C K A O	D P A Z O	B C M R I	A F N W O	G L I H T
Plain	N F I F T	E E N T H	I N F A N	T R Y S E	C T O R A	T O N C E
Cipher	P C K A O	D P A Z O	B C M R I	A F N W O	G L I H T	I W W C U

No. 2

Key	L A T E R	B K K M J	R B T U X	S G E B Q	Y R H H A	T E T U C
Plain	R E Q U E	S T I N F	O R M A T	I O N O F	S I T U A	T I O N I
Cipher	B K K M J	R B T U X	S G E B Q	Y R H H A	T E T U C	N O G T M
Key	N O G T M	L D Q L E	N G B Y E	W D S U H	P U T Z E	H H G D K
Plain	N F I F T	E E N T H	I N F A N	T R Y S E	C T O R A	T O N C E
Cipher	L D Q L E	N G B Y E	W D S U H	P U T Z E	H H G D K	T O D E X

(2) Now let the two cryptograms be superimposed and isomorphisms be sought. They are shown underlined below:

1	T S B J S	M M N R U	L P U I H	J B T X F	I N N R M	D W I Q V
2	B K K M J	R B T U X	S G E B Q	Y R H H A	T E T U C	N O G T M
1	P C K A O	D P A Z O	B C M R I	A F N W O	G L I H T	I W W C U
2	L D Q L E	N G B Y E	W D S U H	P U T Z E	H H G D K	T O D E X

It will be noted that the intervals between isomorphic superimposed pairs show a constant factor of 5, indicating a 5-letter intial key word.

(3) A reconstruction diagram for the pairs beyond the first five letters is established, based upon this interval of 5, and is as follows:

	A	B	C	D	E	F	G	H	I	J	K	L	M	N	O	P	Q	R	S	T	U	V	W	X	Y	Z
1	P	W		N			H		T	Y	D	S	R			L	I	O					F	G		
2	X	R	D			U						H	B	E		G		W					O	P		
3	B	K		I		N	O		G		Q		S	T		W	X	C		H	E		D	R		
4	L	F	E	A				D	B		N	C		P		S	T	U		W			Z	H		Y
5	W	D		T		A	U	Q	H		I		C	B	E	F	G			K	X	M	N	O		

The equivalent sequence A W N B D T K I H Q G U X O E R V M C Y S J L Z P F is established by indirect symmetry; from this, by decimation on the eleventh interval, the HYDRAULIC . . . XZ component is recovered.

(4) It will be noted that the foregoing case, in which the initial key words for the two cryptograms are of the same length, is only a special application of the method set forth in paragraph 44 of Military Cryptanalysis, Part II. But if the key words were of different lengths, the method set forth in paragraph 45 of the text referred to would be applicable. No example is deemed necessary, since no new principles are involved.

SECTION VIII

SOLUTION OF PLAIN-TEXT AUTO-KEY SYSTEMS

30. **Preliminary remarks on plain-text auto-keying.**—*a*. If the cipher alphabets are unknown sequences, plain-text auto-keying gives rise to cryptograms of more intricate character than does cipher-text auto-keying, as has already been stated. As a cryptographic principle it is very commonly encountered as a new and remarkable "invention" of tyros in the cryptographic art. It apparently gives rise to the type of reasoning to which attention has been directed once before and which was then shown to be a popular delusion of the uninitiated. The novice to whom the auto-key principle comes as a brilliant flash of the imagination sees only the apparent impossibility of penetrating a secret which enfolds another secret. His reasoning runs about as follows: "In order to read the cryptogram, the would-be solver must, of course, first know the key; but the key does not become known to the would-be solver until he has read the cryptogram and has thus found the plain text. Since this is reasoning around a circle, the system is indecipherable." How unwarranted such reasoning really is in this case, and how readily the problem is solved, will be demonstrated in the next few paragraphs.

b. A consideration of the mechanics of the plain-text auto-key method discloses that a repetition of n letters in the plain text will produce a repetition of $(n-k)$ letters in the cipher text, where n represents the length of the repetition and k the length of the introductory key. Therefore, when the introductory key consists of a single letter there will be as many repetitions in the cipher text as there are in the plain text, except for true digraphic repetitions, which of course disappear. But on the other hand some "accidental" digraphic repetitions are to be fairly expected, since it can happen that two different plain-text pairs, enciphered by different key letters, will produce identical cipher equivalents. Such accidental repetitions will happen less frequently, of course, in the case of longer polygraphs, so that when repetitions of 4 or more letters are found in the cipher text they may be taken to be true or causal repetitions. It is obvious that in studying repetitions in a cryptogram of this type, when the introductory key is a single letter, a 5-letter repetition in the cipher text, for example, represents a 6-letter word, or sequence repeated in the plain text. When the introductory key is k letters in length then an n-letter repetition represents an $(n+k)$-letter repetition in the plain text.

c. The discussion will, as usual, be divided into two principal cases: (1) when the cipher alphabets are known and (2) when they are unknown. Under each case there may be an introductory key consisting of a single letter, a word, or a short phrase. The single-letter initial key will be treated first.

31. **Solution of plain-text auto-keyed cryptograms when the introductory key is a single letter.**—*a*. Note the following plain-text auto-keyed encipherment of such commonly encountered plain-text words as COMMANDING, BATTALION, and DIVISION, using two identical primary components, in this case direct standard alphabets:

(1)	Key text	. B A T T A L I O N	Key text	. D I V I S I O N	(2)
	Plain text	B A T T A L I O N .	Plain text	D I V I S I O N .	
	Cipher	. B T M T L T W B .	Cipher	. L D D A A W B .	
(3)	Key text	. C O M M A N D I N G	Key text	. C A P T A I N	(4)
	Plain text	C O M M A N D I N G .	Plain text	C A P T A I N .	
	Cipher	. Q A Y M N Q L V T .	Cipher	. C P I T I V .	

These characteristics may be noted:[1]

(1) The cipher equivalent of A_p is the plain-text letter which immediately precedes A_p. (See the two A's in BATTALION, in example 1 above.)

(2) A plain-text sequence of the general formula ABA yields a doublet as the cipher equivalent of the final two letters. (See IVI or ISI in DIVISION, example 2 above.)

(3) Every plain-text trigraph having A_p as its central letter yields a cipher equivalent the last two letters of which are identical with the initial and final letters of the plain-text trigraph. (See MAN in COMMANDING, example 3 above.)

(4) Every plain-text tetragraph having A_p as the initial and the final letter yields a cipher equivalent the second and fourth letters of which are identical with the second and third letters of the plain-text tetragraph, respectively. (See APTA in CAPTAIN, example 4 above; also ATTA in BATTALION, example 1.)

b. (1) From the foregoing characteristics and the fact that a repetition of a sequence of n plain-text letters will yield, in the case of a 1-letter introductory key, a repetition of a sequence of n-l cipher letters, it is obvious that the simplest method of solving this type of cipher is that of the probable word. Indeed, if the system were used for regular traffic it would not be long before the solution would consist merely in referring to lists of cipher equivalents of commonly used words (as found from previous messages) and searching through the messages for these cipher equivalents.

(2) Note how easily the following message can be solved:

B E C J I   B T M T L   T W B P Q   A Y M N Q   H V N E T   W A A L C...

Seeing the sequence BTMTLTWB, which is on the list of equivalents in *a* above (see example 1), the word BATTALION is inserted in proper position. Thus:

B E C J I   B T M T L   T W B P Q . . .
. . . . B   A T T A L   I O N . . . . .

With this as a start, the decipherment may proceed forward or backward with ease. Thus:

B E C J I   B T M T L   T W B P Q   A Y M N Q   H V N E T   W A A L C . . .
E A C H B   A T T A L   I O N C O   M M A N D   E R W I L   L P L A C . . .

c. The foregoing example is based upon the so-called Vigenère method of encipherment ($\Theta_{k/2}=\Theta_{1/1}$; $\Theta_{p/1}=\Theta_{c/2}$). If in encipherment the plain-text letter is sought in the cipher component, its equivalent taken in the plain component ($\Theta_{k/2}=\Theta_{1/1}$; $\Theta_{p/2}=\Theta_{c/1}$), the steps in solution are identical, except that the list of cipher equivalents of probable words must be modified accordingly. For instance, BATTALION will now be enciphered by the sequence................ZTAHLXGZ.

[1] The student is cautioned that the characteristics noted apply only to the case where two identical components are used, with the base letter A.

d. If reversed standard cipher alphabets are used, the word BATTALION will be enciphered by the sequence.................................... BHATPDUB, which also presents idiomorphic characteristics leading to the easy recognition of the word.

e. All the foregoing phenomena are based upon standard alphabets, but when mixed cipher components are used and these have been reconstructed, similar observations may be recorded and the results employed in the solution of additional messages enciphered by the same components.

32. Example of solution by the probable-word method.—*a*. The solution of messages enciphered by unknown mixed components will now be discussed by example. When the primary components are unknown, the observations noted under the preceding subparagraphs are, of course, not applicable; nevertheless solution is not difficult. Given the following three cryptograms, all intercepted on the same day, and therefore suspected of being related:

MESSAGE I

HUFII OCQJJ IVZOZ VPDGO VVVKW
UEWHU UQHUM RZVQR UAKVD NNEZV
GJPGH AYJDR UWNGR YSKBL QVUXN
PHDPR SVKZP PPKGS LLPRV RBHAK
WUAVW YUEZQ XAPQY GPSVS FNRAK
CIFGZ UVCCP DKCWV XTWFM RFKBV
ROQOJ DRUWN GRYSK BL

MESSAGE II

JUFII OCQJJ IVZOZ IBFEJ SUBRJ
SPKTS RZVXT WFMRF QHHFO RFJPD
GOVVV KWUHE NDBDD RHWUN KCMPD
GOVZS ENDBD DRHWU NPPKP EQOY

MESSAGE III

FJUHF FKDEN ALUPZ KQMVB JWVPK
EUBDD RHWUM RHVGP DNCUJ CDZCY
RHUJU FZPQP YQCYH OEQZV XKCQF
TVHNS VCCEJ PEAMP APOEP BHMVJ
UNMHH WKCVG DSWJA EQZBU FFYUE
ZQXAP QYGPA RPZVX CFNRA KCIFG
ZUVCC PDKCO GJWZH APUFZ FVHAV
XMHFF KMYHS TBSKC VRQIJ YCPZH
UHCBM THOFH

b. (1) There are many repetitions, their intervals show no common factor, and a uniliteral frequency distribution does not appear to be monoalphabetic. Plain-text auto-keying is suspected. The simplest assumption to make at the start is that single-letter introductory keys are being used, with the normal Vigenère method of encipherment, and that the plain component is the normal sequence. Attempts to solve any of the messages on the assumption that the cipher component is also the normal sequence being unsuccessful, it is next assumed that the cipher component is a mixed sequence. The 13-letter repetition J D R U W N G R Y S K B L and the 10-letter repetition P D G O V V V K W U are studied intensively. If a single-letter introductory key is being used, then these repetitions involve 14-letter and 11-letter plain-text sequences or words; if the normal Vigenère method of encipherment is in effect $(\Theta_{k/2}=\Theta_{i/1};\ \Theta_{p/1}=\Theta_{c/2})$, then the base letter is A. If the latter is true then a good word which would fit the 13-letter repetition is:

Key	.	R	E	C	O	N	N	A	I	S	S	A	N	C	E
Plain text	R	E	C	O	N	N	A	I	S	S	A	N	C	E	.
Cipher		J	D	R	U	W	N	G	R	Y	S	K	B	L	.

and a good word which would fit the 10-letter repetition is:

Key	.	O	B	S	E	R	V	A	T	I	O	N
Plain text	O	B	S	E	R	V	A	T	I	O	N	.
Cipher		P	D	G	O	V	V	V	K	W	U	.

(2) Inserting, in a mixed component, the values given by these two assumptions yields the following:

Plain	A	B	C	D	E	F	G	H	I	J	K	L	M	N	O	P	Q	R	S	T	U	V	W	X	Y	Z
Cipher	R		A		J		S	T	I	N	G	B	C			K	L					V	W		Y	
	E		D		U													O	P							

(3) It is a simple matter to combine these two partial cipher components into a single sequence, and the two components are as follows:

Plain	A	B	C	D	E	F	G	H	I	J	K	L	M	N	O	P	Q	R	S	T	U	V	W	X	Y	Z
Cipher	R	E	A	D	J	U	S	T	I	N	G	B	C	F	H	K	L	M	O	P	Q	V	W	X	Y	Z

(4) With the primary components at hand, solution of the messages is now an easy matter.

c. The foregoing example uses an unknown mixed cipher component sliding against what was first assumed (and later proved) to be the normal direct sequence. When both primary components are unknown mixed sequences but are identical, solution is more difficult, naturally, because the results of assuming values for repeated sequences cannot be proved and established so quickly as in the foregoing example. Nevertheless, the general method indicated, and the application of the principles of indirect symmetry will lead to solution, if there is a fair amount of text available for study. When an introductory key of several letters is used, repetitions are much reduced and the problem becomes still more difficult but by no means insurmountable. Space forbids a detailed treatment of the method of solving these cases but it is believed that the student is in a position to develop these methods and to experiment with them at his leisure.

33. Concluding remarks on the solution of auto-key systems.—*a.* The type of solution elucidated in the preceding paragraph is based upon the successful application of the probable-word method. But sometimes the latter method fails because the commonly expected words may not be present after all. Hence, other principles and methods may be useful. Some of these methods, useful in special cases, are almost mechanical in their nature. Extension of the basic principles involved may lead to rather far-reaching complexities. However, because these methods are applicable only to somewhat special situations, and because they are somewhat involved they will be omitted from the text proper and placed in Appendix 1. The student who is especially interested in these cases may consult that appendix at his leisure.

b. It is thought that sufficient attention has been devoted to the solution of both cipher-text and plain-text auto-key systems to have demonstrated to the student that these cryptographic methods have serious weaknesses which exclude them from practical usage in military cryptography. Besides being comparatively slow and subject to error, they are rather easily solvable, even when unknown cipher alphabets are employed.

c. In both systems there are characteristics which permit of identifying a cryptogram as belonging to this class of substitution. Both cases will show repetitions in the cipher text. In cipher-text auto-keying there will be far fewer repetitions than in the original plain text, especially when introductory keys of more than 1-letter in length are employed. In plain-text auto-keying there will be nearly as many repetitions in the cipher text as in the original plain text unless long introductory keys are used. In either system the repetitions will show no constancy as regards intervals between them, and a uniliteral frequency distribution will show such messages to be polyalphabetic in nature. Cipher-text auto-keying may be distinguished from plain-text auto-keying by the appearance of the frequency distribution of the second member of sets of two letters separated by the length of the introductory key (see par. 26*b*, *c*). In the case of cipher-text auto-keying these frequency distributions will be monoalphabetic in nature; in plain-text auto-keying such frequency distributions will not show monoalphabetic characteristics.

SECTION IX

METHODS OF LENGTHENING OR EXTENDING THE KEY

34. Preliminary remarks.—In paragraph 1*b* of this text it was stated that two procedures suggest themselves for eliminating the weaknesses introduced by periodicity of the type produced by simple, repeating-key methods. The first of these, when studied, embraced some of the very simple methods of suppressing or destroying periodicity, by such devices as interrupting the key and using variable-length groupings of plain text. It was demonstrated that subterfuges of this simple nature are inadequate to eliminate the weaknesses referred to, and must be discarded in any system intended to afford real security. The other alternative suggested in paragraph 1*b* therefore remains now to be investigated, viz, that of lengthening the keys to a point where there would seem to be an insufficient amount of text to enable the cryptanalyst to solve the traffic. Attempts toward this end usually consist in extending the key to such a length that the enemy cryptanalysts will have only a very limited number of periods to work with. The key may, indeed, be lengthened to a point where it becomes as long as, or longer than, the text to be enciphered, so that the key is used only once.

35. Extended and nonrepeating keys.—*a*. It is obvious that one of the simplest methods of lengthening the key to a message is to use a long phrase or even a complete sentence, provided it is not too long to remember. In addition to the difficulties that would be encountered in practical military cryptography in selecting long mnemonic phrases and sentences which would have to be imparted to many clerks, there is the fact that the probable-word method of solution still remains as a powerful tool in the hands of enemy cryptanalysts. And if only a word or two of the key can be reconstructed as a result of a fortunate assumption, it is obvious that the enemy cryptanalysts could readily guess the entire key from a fragment thereof, since any long phrase or sentence which is selected because it can easily be remembered is likely to be well known to many people.

b. There are, however, more or less simple methods of employing a short mnemonic key in order to produce a much longer key. Basically, any method of transposition applied to a single alphabetic sequence repeated several times will yield a fairly long key, which, moreover, has the advantage of being unintelligible and thus approaching a random selection of letters. For example, a numerical key may be derived from a word or a short phrase; this numerical key may then be applied as a columnar-transposition key for a rectangle within which the normal alphabet has been repeated a previously agreed upon number of times in a normal (left to right) or prearranged manner. The letters when transcribed from the transposition rectangle then become the successive letters for enciphering the plain text, using any desired type of primary components. Or, if a single transposition is not thought to be sufficiently secure, a double transposition will yield a still more mixed up sequence of key letters. Other types of transposition may be employed for the purpose, including various kinds of geometric figures. Also, a non-transposition method of lengthening the keying sequence and at the same time introducing an irregularity, such as aperiodic interruption has already been described (see par. 18).

c. Another method of developing a long key from a short mnemonic one is that shown below. Given the keyword CHRISTMAS, a numerical sequence is first derived and then one writes down successive sections of this numerical key, these sections terminating with the successive numbers 1, 2, 3, . . . of the numerical key. Thus:

```
                Mnemonic key.......  C H R I S T M A S
                Numerical key......  2-3-6-4-7-9-5-1-8

                                 1|2|  3|      4|            5|    6|
Extended key.......  C H R I S T M A|C|C H|C H R I|C H R I S T M|C H R|
                             7|                8|          9|
                     C H R I S|C H R I S T M A S|C H R I S T|
```

Thus the original key of only 9 letters is expanded to one of 45 letters (1+2+3+ . . . +9=45). The longer key is also an interrupted key of the type noted under paragraph 17, but if the message is long enough to require several repetitions of the expanded key the encipherment becomes periodic and can be handled by the usual methods employed in solving repeating-key ciphers. If the basic key is fairly long, so that the expanded key becomes a quite lengthy sequence, then the message or messages may be handled in the manner explained in paragraph 20.

d. Another method of producing a rather long sequence of digits for keying purposes from a single key number is to select a number whose reciprocal when converted by actual division into its equivalent decimal yields a long series of digits. For example the reciprocal of 49, or 1/49, yields a sequence of 42 digits beginning .02040815 . . . . Such a number, coupled with a key word like CHRISTMAS, could be used for interrupted keying, the successive cipher alphabets being used for enciphering as many letters as are indicated by the successive digits. In the case of the example cited, the first digit is 0; hence the C alphabet would not be used. The next digit is 2; the H alphabet would be used for enciphering the first and second letters. The third digit is again 0; the R alphabet would not be used. The fourth digit is 4; the I alphabet would be used for enciphering the third, fourth, fifth, and sixth letters, and so on.

36. Other systems employing lengthy keying sequences.—*a. The so-called "running-key" system.*—To be mentioned in connection with this subject of extensive or lengthy keys is the cipher system known as the running-key, continuous-key, or nonrepeating-key system, in which the key consists of a sequence of elements which never repeats no matter how long the message to be enciphered happens to be. The most common and most practical source of such a key is that in which the plain text of a previously agreed-upon book serves as the source for successive key letters for encipherment.[1] The solution of this type of cipher, an accomplishment which was once thought impossible, presents some interesting phases and will be considered shortly. At this point it is merely desired to indicate that according to the running-key system the key for an individual message may be as long as the message and never repeat; but if a large group of correspondents employ the same key sequence, it may happen that there will be several messages in the same key and they will all begin with the same initial key letter; or, there will be several which will "overlap" one another with respect to the key, that is, they begin at different initial points in the keying sequence but one message soon overtakes the other, so that from that point forward all subsequent letters in both messages are enciphered by the same sequence of key letters.

---

[1] Sec. IX, *Advanced Military Cryptography.* See also footnote 8, page 71 of this text.

b. The so-called progressive-alphabet system.—In the so-called progressive-alphabet system the basic principle is quite simple. Two or more primary elements are arranged or provided for according to a key which may be varied from time to time; the interaction of the primary elements results in making available for cryptographic purposes a set of cipher alphabets; all the latter are employed in a fixed sequence or progression; hence the designation progressive-alphabet system. If the number of alphabets available for such use is rather small, and if the text to be enciphered is much longer than the sequence of alphabets, then the system reduces to a periodic method. But if the number of alphabets is large, so that the sequence is not repeated, then of course, the cryptographic text will exhibit no periodic phenomena.

c. The series of cipher alphabets in such a system constitutes a keying sequence. Once set up, often the only remaining element in the key for a specific message is the starting point in the sequence, that is, the initial cipher alphabet employed in enciphering a given message. If this keying sequence must be employed by a large group of correspondents, and if all messages employ the same starting point in the keying sequence, obviously the cryptograms may simply be superimposed without any preliminary testing to ascertain proper points for superimposition. The student has already been shown how cases of this sort may be solved. However, if messages are enciphered with varying starting points, the matter of superimposing them properly takes on a different aspect. This will soon be treated in detail.

d. The respective cipher alphabets constituting the entire complement of alphabets may be employed in a simple progression, that is, consecutively from a preselected initial point; or, they may be employed according to other types of progression. For example, if the system comprises 100 alphabets one might use them in the sequence 1, 3, 5, 7, . . . ; or 1, 4, 7, 10, . . . ; or irregular types of skipping may be employed.

e. In addition to the foregoing, there are, of course, a great many mechanical methods of producing a long key, such as those employed in mechanical or electrical cipher machines. In most cases these methods depend upon the interaction of two or more short, primary keys which jointly produce a single, much longer, secondary or resultant key. (See par. 4.) Only brief reference can be made at this point in the cryptanalytic studies to cases of this kind. A detailed treatment of complex examples would require much time and space so that it will be reserved for subsequent texts.

f. Finally, there must be mentioned certain devices in which, as in encipherment by the auto-key method, the text itself serves to produce the variation in cipher equivalents, by controlling the selection of secondary alphabets, or by influencing or determining the sequence with which they will be employed. Naturally, in such cases the key is automatically extended to a point where it coincides in length with that of the text. An excellent example of such a device is that known as the Wheatstone, the solution of which will be described in its proper place.[2] Some writers classify and treat this method as well as auto-key methods as forms of the running-key system but the present author prefers to consider the latter as being radically different in principle from the former types, because in the true running-key system the key is wholly external to and independent of text being enciphered. This is hardly true of auto-key systems or of systems such as the Wheatstone mentioned herein.

[2] See Sec. XII, *Advanced Military Cryptography*.

Section X

GENERAL PRINCIPLES UNDERLYING SOLUTION OF SYSTEMS EMPLOYING LONG OR CONTINUOUS KEYS

37. Solution when the primary components are known sequences.—*a.* As usual, the solution of cases involving long or continuous keys will be treated under two headings: First, when the primary components are known sequences; second, when these elements are wholly unknown or partially unknown.

b. Since the essential purpose in using long keys is to prevent the formation of repetitive cycles within the text, it is obvious that in the case of very long keying sequences the cryptanalyst is not going to be able to take the text and break it up into a number of small cycles which will permit the establishment of monoalphabetic frequency distributions that can readily be solved, an end which he can attain all the more readily if to begin with he knows the primary sequences. But, there nearly always remains the cryptanalyst's last resort: the probable-word method. Inasmuch as this method is applicable to most of these cases, even to that of the running-key system, which perhaps represents the furthest extension of the principle of long keying sequences, an example using a cryptogram of the latter type will be studied.

38. Solution of a running-key cipher when an unknown but intelligible key sequence is used and the primary components are known.—*a.* In paragraph 36*a* mention was made of the so-called running-key, continuous-key, or nonrepeating-key system, in which the plain text of a previously agreed-upon book serves as the source for successive key letters for encipherment. Since the running-key system is entirely aperiodic, and the cipher text can therefore not be arranged in superimposed short cycles, as in the case of the repeating-key system, it would appear on first consideration to be "indecipherable" without the key.[1] But if the student will bear in mind that one of the practical methods of solving a repeating-key cipher is that of the probable word,[2] he will immediately see that the latter method can also be applied in solving this type of nonrepeating-key system. The essence of the matter is this: The cryptanalyst may assume the presence of a probable word in the text of the message; if he knows the primary components involved, and if the assumed word actually exists in the message, he can locate it by checking against the key, *since the latter is intelligible text.* Or, he may assume the presence of a probable word or even of a phrase such as "to the," "of the," etc., in the key text and check his assumption against the text of the message. Once he has forced

[1] At one time, indeed, this view was current among certain cryptographers, who thought that the principle of factoring the intervals between repetitions in the case of the repeating-key cipher formed the basis for the only possible method of solving the latter type of system. Since, according to this erroneous idea, factoring cannot be applied in the case of the running-key system (using a book as the key), therefore solution was considered to be impossible. How far this idea is from the truth will presently be seen. In this same connection see also footnote 8, page 71.

[2] See *Military Cryptanalysis*, Part II, par. 25.

such an entering wedge into either the message or the key, he may build upon this foundation by extending his assumptions for text alternately in the key and in the message, thus gradually reconstructing both. For example, given a cryptogram containing the sequence . . . HVGGLOWBESLTR . . . , suppose he assumes the presence of the phrase THAT THE in the key text and finds a place in the plain text where this yields MMUNITI . Thus, using reversed standard cipher alphabets:

Assumed key text	. . . T H A T T H E . . .
Cipher text	. . . H V G G L O W B E S L T R . . .
Resultant plain text	. . . M M U N I T I . . .

This suggests the word AMMUNITION. The ON in the cipher text then yields PR as the beginning of the word after THE in the key text. Thus:

Assumed key text	. . . T H A T T H E P R . . .
Cipher text	. . . H V G G L O W B E S L T R . . .
Resultant plain text	. . . M M U N I T I O N . . .

PR must be followed by a vowel, with O the most likely candidate. He finds that O yields W in the plain text, which suggests the word WILL. The latter then yields OTEC in the key, making the latter read THAT THE PROTEC . . . . Thus:

Assumed key text	. . . T H A T T H E P R O T E C . . .
Cipher text	. . . H V G G L O W B E S L T R . . .
Resultant plain text	. . . M M U N I T I O N W I L L . . .

This suggests the words PROTECTION, PROTECTIVE, PROTECTING, etc. Thus extending one text a few letters serves to "coerce" a few more letters out of the other, somewhat as in the case of two boys who are running approximately abreast in a race; as soon as one boy gets a bit ahead the spirit of competition causes the other to overtake and pass the first one; then the latter puts forth a little more effort, overtakes and passes the second boy. Thus the boys alternate in overtaking and passing each other until the race is run. The only point in which the simile fails is that while the boys usually run forward all the time, that is, in a single direction, the cryptanalyst is free to work in two directions—forward and backward from an internal point in the message. He may, in the case of the example cited above, continue his building-up process by adding A to the front of MMUNITI as well as ON to the rear. If he reaches the end of his resources on one end, there remains the other end for experimentation. He is certainly unlucky if both ends terminate in complete words both for the message and for the key, leaving him without a single clue to the next word in either, and forcing him to a more intensive use of his imagination, guided only by the context.

b. In the foregoing illustration the cryptanalyst is assumed to have only one message available for his experimentation. But if he has two or more messages which either begin at identical initial points with reference to the key, or overlap one another with respect to the key, the reconstruction process described above is, of course, much easier and is accomplished much more quickly. *For if the messages have been correctly superimposed with reference to the key text, the addition of one or two letters to the key yields suggestions for the assumption of words in several messages.* The latter lead to the addition of several letters to the key, and so on, in an ever-widening circle of ideas for further assumptions, since as the process continues the context affords more and more of a basis for the work.

c. Of course, if sufficient of the key text is reconstructed, the cryptanalyst might identify the book that is being used for the key, and if available, his subsequent labors are very much simplified.

d. All the foregoing is, however, dependent not only upon the use of an intelligible text as the keying text but also upon having a knowledge of the primary components or cipher alphabets employed in the encipherment. Even if the primary components are differently mixed sequences, so long as they are known sequences, the procedure is quite obvious in view of the foregoing explanation. The training the student has already had is believed sufficient to indicate to him the procedure he may follow in that solution, and no further details will here be given in respect to such cases. But what if the primary components are not known sequences? This contingency will be treated presently.

39. Solution of a progressive-alphabet cipher when the cipher alphabets are known.—*a.* Taking a very simple case, suppose the interacting elements referred to in paragraph 36*b* consist merely of two primary cipher components which slide against each other to produce a set of 26 secondary cipher alphabets, and suppose that the simplest type of progression is used, *viz*, the cipher alphabets are employed one after the other consecutively. Beginning at an initial juxtaposition, producing say, alphabet 1, the subsequent secondary alphabets are in the sequence 2, 3, . . . 26, 1, 2, 3, . . ., and so on. If a different initial juxtaposition is used, say alphabet 10 is the first one, the sequence is exactly the same as before, only beginning at a different point.

b. Suppose the two primary components are based upon the keyword HYDRAULIC. A message is to be enciphered, beginning with alphabet 1. Thus:

Plain component..........H Y D R A U L I C B E F G J K M N O P Q S T V W X Z H Y D . . .
Cipher component.........H Y D R A U L I C B E F G J K M N O P Q S T V W X Z

Letter No.	1	2	3	4	5	6	7	8	9	10	11	12	13	14	15	16	17	18	19	20	21
Alphabet No.	1	2	3	4	5	6	7	8	9	10	11	12	13	14	15	16	17	18	19	20	21
Plain text	E	N	E	M	Y	H	A	S	P	L	A	C	E	D	H	E	A	V	Y	I	N
Cipher text	E	O	G	P	U	U	E	Y	H	M	K	Q	V	M	K	Z	S	J	Q	H	E

Letter No.	22	23	24	25	26	27	28	29	30	31	32	33	34	35	36	37	38	39
Alphabet No.	22	23	24	25	26	1	2	3	4	5	6	7	8	9	10	11	12	13
Plain text	T	E	R	D	I	C	T	I	O	N	F	I	R	E	U	P	O	N
Cipher text	N	L	H	H	L	C	V	B	S	S	N	J	E	P	K	D	D	D

Letter No.	40	41	42	43	44	45	46	47	48	49	50	51	52	53
Alphabet No.	14	15	16	17	18	19	20	21	22	23	24	25	26	1
Plain text	Z	A	N	E	S	V	I	L	L	E	R	O	A	D
Cipher text	G	P	U	H	F	K	H	H	Y	L	H	M	R	D

c. This method reduces to a periodic system involving 26 secondary cipher alphabets and the latter are used in simple progression. It is obvious therefore that the 1st, 27th, 53d, . . . letters are in the 1st alphabet; the 2d, 28th, 54th, . . . letters are in the 2d alphabet, and so on.

d. To solve such a cryptogram, knowing the two primary components, is hardly a problem at all. The only element lacking is a knowledge of the starting point. But this is not necessary, for merely by completing the plain-component sequences and examining the diagonals of the diagram, the plain text becomes evident. For example, given the following: HIDCT EHUXI. Completing the plain-component sequences initiated by the successive cipher letters, the plain text, E N E M Y M A C H I . . . is seen to come out in successive steps upward in Figure 10. Had the cipher component been shifted in the opposite direction in encipherment, the steps would have been downward instead of upward. If the sliding strips had been set up according to the sequence of cipher letters but on a diagonal, then, of course, the plain-text letters would have reappeared on one generatrix.

e. The student will understand what simple modifications in procedure would be required in case the two primary components were different mixed sequences. But what if the primary components are not known sequences? How does the cryptanalyst proceed in that case?

```
H I D C T E H U X L
Y C R B V F Y L Z I
D B A E W G D I H C
R E U F X J R C Y B
A F L G Z K A B D E
U G I J H M U E R F
L J C K Y N L F A G
I K B M D O I G U J
C M E N R P C J L K
B N F O A Q B K I M
E O G P U S E M C N
F P J Q L T F N B O
G Q K S I V G O E P
```

FIGURE 10.

40. General solution for ciphers involving a long-keying sequence of fixed length and composition.—*a*. It is obvious, as stated at a previous point, that no matter how the keying sequence is derived, *if all the correspondents employ the same key, or if this key is used many times by a single office, and if it always begins at the same point, the various messages may simply be superimposed.* Thus, their respective 1st, 2d, 3rd, . . . letters will all fall within columns which have been enciphered by the 1st, 2d, 3rd, . . . key letters. If there is a sufficient number of messages, solution then becomes possible by frequency analysis of the successive columns—no matter how long the keying sequence may be, and regardless of whether the keying sequence constitutes intelligible text or is a purely random sequence of letters. This method of solution by superimposition has already been outlined in paragraph 20 and no further reference to it need here be made.

b. But now suppose that the keying sequence does not always begin at the same point for all messages. Suppose the several correspondents are able to select at will *any* point in the keying sequence as the point of departure in encipherment. Thus, such a keying sequence, if regarded as partaking of the nature of a circle, will afford as many possible starting points as there are letters or characters in that sequence. Now if there are no external indications or *indicators* [2] in the cryptograms pertaining to such a system, such as would afford enemy cryptanalysts direct and definite information with regard to the initial keying element for each cryptogram, then it would seem as though the superimposition of messages (to bring letters enciphered by the same cipher alphabets within the same columns) would be difficult or impossible, and therefore that attempts at solution are blocked at their very beginning. This, however, is not the end of the story. For suppose two of the messages have in common only one polygraph, say of 5 letters; these two messages may be juxtaposed so as to bring these repetitions into superimposition. Thus, the possession of this long polygraph in common serves to "tie" these two messages together or to "interlock" them. Then, suppose a shorter polygraph, say of 4 letters, is possessed in common by one of these two messages and a third message; this will serve to tie in the latter with the first two. Extension of this process, including the data from shorter repetitions of trigraphs and digraphs, will serve to assemble a whole set of such messages in proper superimposition. Therefore, the first step is to examine all the messages for repetitions.

[2] Indicators play an important rôle in practical cryptography. An indicator is a symbol (consisting of a letter, group of letters, a figure or a group of figures) which indicates the specific key used under the general cryptographic system, or it may indicate which one of a number of general systems has been used, or it may indicate both.

c. When such repetitions are found, and if there are plenty of them so that assumptions for probable words are easy to make, it is clear that the correct assumptions will enable the cryptanalyst to set up plain-cipher equivalencies which will make it possible to reconstruct the primary components. Depending upon the type used, the principles of direct or indirect symmetry of position will be very useful in this process.

d. But if it happens that there are no polygraphs by means of which two or more messages may be tied together and properly superimposed, the simple methods mentioned in subparagraphs *a–c* cannot here be applied. However, although the road toward a solution seems to be blocked rather effectively, there is a detour which presents rather interesting vistas. The latter are really of such importance in cryptanalysis as to warrant detailed treatment.

SECTION XI

THE "COINCIDENCE" OR "κ" TEST

41. The basic theory of the coincidence or κ (kappa) test.—*a.* In Appendix 2 of the preceding text [1] certain simple applications of the theory of probability were presented for the student's consideration, by way of pointing out to him the important role which certain phases of that branch of mathematics play in cryptanalysis. Reference was there made to the subject *of coincidences* and its significance in connection with the study of repetitions in cryptograms. In this section the matter will be pursued a few steps further.

b. In the appendix referred to, it was shown that the probability of monographic coincidence (1) in random text employing a 26-letter alphabet is .0385; (2) in English telegraphic plain text, .0667. These two parameters were represented by the symbols κ_r and κ_p, respectively. The important role which these values play in a certain cryptanalytic test will now be explained.

c. One of the most important techniques in cryptanalytics is that known as *applying the coincidence or "kappa test."* This test is useful for several cryptanalytic purposes and one of the most important of them is to ascertain when two or more sequences of letters are correctly superimposed. By the word "correct" in this case is merely meant that the sequences are so arranged relative to one another as to facilitate or make possible a solution. The test has for its theoretical basis the following circumstances:

(1) If any two rather lengthy sequences of characters are superimposed, it will be found, on examining both members of the successive pairs of letters brought into vertical juxtaposition, that *in a certain number of cases the two superimposed letters will coincide.*

(2) If both sequences of letters constitute random text (of a 26-letter alphabet), there will be about 38 or 39 such cases of coincidence per thousand pairs examined. This, of course, is because $\kappa_r = .0385$.

(3) If both sequences of letters constitute plain text, there will be about 66 or 67 such cases of coincidence per thousand pairs examined. This is because κ_p is .0667.

(4) If the superimposed sequences are wholly monoalphabetic encipherments of plain text by the same cipher alphabet, there will still be about 66 or 67 cases of coincidence in each 1,000 cases examined, because in monoalphabetic substitution there is a fixed or unvarying relation between plain-text letters and cipher letters so that for statistical purposes monoalphabetic cipher text behaves just the same as if it were normal plain text.

(5) Even if the two superimposed sequences are not monoalphabetically enciphered texts, but are polyalphabetic in character, there will still be about 66 or 67 cases of identity between superimposed letters per thousand cases examined, *provided the two sequences really belong to the same cryptographic system and are superimposed at the proper point with respect to the keying sequence.* The reasons for this will be set forth in the succeeding subparagraphs.

[1] *Military Cryptanalysis, Part II.* It is recommended that the student refresh his memory by reviewing that appendix.

(6) Consider the two messages below. They have been enciphered polyalphabetically by the same two primary components sliding against each other. The two messages use the same keying sequence, beginning at the same initial point in that sequence. Consequently, the two messages are identically enciphered, letter for letter, and the only differences between them are those occasioned by differences in plain text.

No. 1	Alphabets	16	21	13	5	6	4	17	19	21	21	2	6	3	6	13	13	1	7	12	6
	Plain text	W	H	E	N	I	N	T	H	E	C	O	U	R	S	E	L	O	N	G	M . . .
	Cipher	E	Q	N	B	T	F	Y	R	C	X	X	L	Q	J	N	Z	O	Y	A	W . . .
No. 2	Alphabets	16	21	13	5	6	4	17	19	21	21	2	6	3	6	13	13	1	7	12	6
	Plain text	T	H	E	G	E	N	E	R	A	L	A	B	S	O	L	U	T	E	L	Y . . .
	Cipher	P	Q	N	T	U	F	B	W	D	J	L	Q	H	Y	Z	P	T	M	Q	I . . .

Note, now, that (a) in every case in which two superimposed cipher letters are the same, the plain-text letters are identical and (b) in every case in which two superimposed cipher letters are different, the plain-text letters are different. In such a system, even though the cipher alphabet changes from letter to letter, the number of cases of identity or coincidence in the two members of a pair of superimposed cipher letters will still be about 66 or 67 per thousand cases examined, *because the two members of each pair of superimposed letters are in the same cipher alphabet and it has been seen in (4) that in monoalphabetic cipher text* κ *is the same as for plain text,*[2] *viz, .0667.* The two messages may here be said to be superimposed "correctly," that is, brought into proper juxtaposition with respect to the keying sequence.

(7) But now suppose the same two messages are superimposed "incorrectly," that is, they are no longer in proper juxtaposition with respect to the keying sequence. Thus:

No. 1	Alphabets	16	21	13	5	6	4	17	19	21	21	2	6	3	5	13	13	1	7	12	
	Plain text	W	H	E	N	I	N	T	H	E	C	O	U	R	S	E	L	O	N	G . . .	
	Cipher	E	Q	N	B	T	F	Y	R	C	X	X	L	Q	J	N	Z	O	Y	A . . .	
No. 2	Alphabets		16	21	13	5	6	4	17	19	21	21	2	6	3	6	13	13	1	7	
	Plain text		T	H	E	G	E	N	E	R	A	L	A	B	S	O	L	U	T	E . . .	
	Cipher		P	Q	N	T	U	F	B	W	D	J	L	Q	H	Y	Z	P	T	M . . .	

It is evident that the two members of every pair of superimposed letters are no longer in the same cipher alphabet, and therefore, if two superimposed cipher letters *are* identical this is merely an "accident," for now there is no basic or general cause for the similarity, such as is true in the case of a correct superimposition. The similarity, if present, is, as already stated, due to chance and the number of such cases of similarity should be about the same as though the two cipher letters were drawn at random from random text, in which $\kappa_r=.0385$. It is no longer true that (a) in every case in which two superimposed cipher letters are the same, the plain-text letters are identical, or (b) in every case in which two superimposed cipher letters are different, the plain-text letters are different. Note, for example, that the superimposed T_c's represent two different plain-text letters and that the S_p of the word COURSE in the first message gives J_c while the S of the word ABSOLUTELY in the second message gives H_c. Thus, it becomes clear that in an incorrect superimposition two different plain-text letters enciphered by two different alphabets may "by chance" produce identical cipher letters, which on superimposition yield a

[2] The fact that in this case each monoalphabet contains but two letters does not affect the theoretical value of κ; and whether the actual number of coincidences agrees closely with the expected number based upon $\kappa=$ .0667 depends upon the lengths of the two superimposed sequences.

coincidence having no external indications as to dissimilarity in plain-text equivalents. Hence, if there are no other factors which enter into the matter and which might operate to distort the results to be expected from the operation of the basic factor, the expected number of cases of identical cipher letters brought together by an incorrect superimposition will be determined by the value $\kappa_r=.0385$.

(8) But now note also that in the foregoing incorrect superimposition there are two Z_c's and that they represent the same plain-text letter L. This is occasioned by the fact that the plain-text messages happened to have L's in just those two places and that the cipher alphabet happened to be the same both times. Hence, it becomes clear that the same cipher alphabet brought into play twice may "by chance" happen to encipher the same plain-text letter both times, thus producing identical cipher letters. In some systems this source of identity in superimposed cipher letters is of little importance, in other systems, it may materially affect the actual number of coincidences. For instance, if a system is such that it produces a long secondary keying cycle composed of repetitions of short primary keying cycles, an incorrect superimposition of two cryptograms may bring into juxtaposition many of these short cycles, with the result that the actual number of cases of identical superimposed cipher letters is much greater than the expected number based upon $\kappa_r=.0385$. Thus, this source for the production of identical cipher letters in an incorrect superimposition operates to increase the number of cases to be expected from the fundamental constant $\kappa_r=.0385$.

(9) In some systems, where nonrelated cipher alphabets are employed, it may happen that two identical plain-text letters may be enciphered by two different cipher alphabets which, "by chance," have the same equivalent for the plain-text letter concerned. This is, however, a function of the particular cryptographic system and can be taken into account when the nature of the system is known.

(10) In general, then, it may be said that in the case of a correct superimposition the probability of identity or coincidence in superimposed cipher letters is .0667; in the case of an incorrect superimposition, the probability is at least .0385 and may be somewhat greater, depending upon special circumstances. The foregoing situation and facts make possible what has been referred to as the "coincidence test." Since this test uses the constant κ, it is also called the "kappa test."

d. The way in which the coincidence test may be applied will now be explained. The statement that $\kappa_p=.0667$ means that in 1,000 cases where two letters are drawn at random from a large volume of plain text, there will be about 66 or 67 cases in which the two letters coincide, that is, are identical. Nothing is specified as to what the two letters shall be; they may be two Z's or they may be two E's. This constant, .0667, really denotes a percentage: If many *comparisons* of single letters are made, the letters being drawn *at random* from among those constituting a large volume of plain text, 6.67 percent of these comparisons made will yield coincidences. So, if 2,000 such comparisons are made, the theory indicates that there should be about $.0667\times2{,}000=133$ coincidences; if there is sufficient text to permit of making 20,000 comparisons, there should be about 1,334 coincidences, and so on.

e. Another way of handling the matter is to find the ratio of the observed number of coincidences to the total number of cases in which the event in question might possibly occur, i. e., the total number of comparisons of superimposed letters. When this ratio is closer to .0667 than it is to .0385 the correct superimposition has been ascertained. This is true because in the case of a correct superimposition both members of each pair of superimposed letters actually belong to the same monoalphabet and therefore the probability of their coinciding is .0667; whereas in the case of an incorrect superimposition the members of each pair of superimposed letters belong, as a general rule, to different monoalphabets [3], and therefore the probability of their coinciding is nearer .0385 than .0667.

f. From the foregoing, it becomes clear that the kappa test involves ascertaining the total number of comparisons that can be made in a given case, as well as ascertaining the actual number of coincidences in the case under consideration. When only two messages are superimposed, this is easy: The total number of comparisons that can be made is the same as the number of superimposed pairs of letters. But when more than two messages are superimposed in a *superimposition diagram* it is necessary to make a simple calculation, based upon the fact that n letters yield $\frac{n(n-1)}{2}$ pairs or comparisons, where n is the number of letters in the column.[4] For example, in the case of a column of 3 letters, there are $\frac{3\times 2}{2}=3$ comparisons. This can be checked by noting that the 1st letter in the column may be compared with the 2d, the 2d with the 3d, and the 1st with the 3d, making 3 comparisons in all. The number of comparisons per column times the number of columns in the superimposition diagram of letters gives the total number of comparisons. The extension of this reasoning to the case where a superimposition diagram has columns of various lengths is quite obvious: one merely adds together the number of comparisons for columns of different lengths to obtain a grand total. For convenience, the following brief table is given:

Number of letters in column	Number of comparisons	Number of letters in column	Number of comparisons	Number of letters in column	Number of comparisons
2	1	11	55	21	210
3	3	12	66	22	231
4	6	13	78	23	253
5	10	14	91	24	276
6	15	15	105	25	300
7	21	16	120	26	325
8	28	17	136	27	351
9	36	18	153	28	378
10	45	19	171	29	406
		20	190	30	435

g. In ascertaining the number of coincidences in the case of a column containing several letters, it is again necessary to use the formula $\frac{n(n-1)}{2}$, only in this case n is the number of identical letters in the column. The reasoning, of course, is the same as before. The total

[3] The qualifying phrase "as a general rule" is intended to cover any distortion in results occasioned by the presence of an unusual number of those cases of coincidence described under subpar. *c* (8) and (9).

[4] This has already been encountered (footnote 3, Appendix 2, *Military Cryptanalysis, Part II*). It is merely a special case under the general formula for ascertaining the number of combinations that may be made of n different things taken r at a time, which is ${}_nC_r=\frac{n!}{r!(n-r)!}$. In studying coincidences by the method indicated, since only two letters are compared at a time, r is always 2; hence the expression $\frac{n!}{r!(n-r)!}$, which is the same as $\frac{n(n-1)(n-2)!}{2(n-2)!}$, becomes by cancellation of $(n-2)!$, reduced to $\frac{n(n-1)}{2}$.

462861 O - 42 - 5

number of coincidences is the sum of the number of coincidences for each case of identity. For example, in the column shown at the side, containing 10 letters, there are 3 B's, 2 C's, 4 K's, and 1 Z. The 3 B's yield 3 coincidences, the 2 C's yield 1 coincidence, and the 4 K's yield 6 coincidences. The sum of 3+1+6 makes a total of 10 coincidences in 45 comparisons.

C
K
B
K
Z
K
C
B
B
K

42. General procedure to be followed in making the κ test.—*a*. The steps in applying the foregoing principles to an actual case will now be described. Suppose several messages enciphered by the same keying sequence but each beginning at a different point in that sequence are to be solved. The indicated method of solution is that of superimposition, the problem being to determine just where the respective messages are to be superimposed so that the cipher text within the respective columns formed by the superimposed messages will be monoalphabetic. From what has been indicated above, it will be understood that the various messages may be shifted relative to one another to many different points of superimposition, there being but one correct superimposition for each message with respect to all the others. First, all the messages are numbered according to their lengths, the longest being assigned the number 1. Commencing with messages 1 and 2, and keeping number 1 in a fixed position, message 2 is placed under it so that the initial letters of the two messages coincide. Then the two letters forming the successive pairs of superimposed letters are examined and the total number of cases in which the superimposed letters are identical is noted, this giving the observed number of coincidences. Next, the total number of superimposed pairs is ascertained, and the latter is multiplied by .0667 to find the expected number of coincidences. If the observed number of coincidences is considerably below the expected number, or if the ratio of the observed number of coincidences to the total number of comparisons is nearer .0385 than .0667, the superimposition is incorrect and message 2 is shifted to the next superimposition, that is, so that its first letter is under the second of message 1. Again the observed number of coincidences is ascertained and is compared with the expected number. Thus, by shifting message 2 one space at a time (to the right or left relative to message 1) the coincidence test finally should indicate the proper relative positions of the two messages. When the correct point of superimposition is reached the cryptanalyst is rarely left in doubt, for the results are sometimes quite startling. After messages 1 and 2 have been properly superimposed, message 3 is tested first against messages 1 and 2 separately, and then against the same two messages combined at their correct superimposition.[5] Thus message 3 is shifted a step each time until its correct position with respect to messages 1 and 2 has been found. Then message 4 is taken and its proper point of superimposition with respect to messages 1, 2, and 3 is ascertained. The process is continued in this manner until the correct points of superimposition for all the messages have been found. It is obvious that as messages are added to the superimposition diagram, the determination of correct points of superimposition for subsequent messages becomes progressively more certain and therefore easier.

b. In the foregoing procedure it is noted that there is necessity for repeated displacement of one message against another or other messages. Therefore, it is advisable to transcribe the messages on long strips of cross-section paper, joining sections accurately if several such strips are necessary to accommodate a long message. Thus, a message once so transcribed can be shifted to various points of superimposition relative to another such message, without repeatedly rewriting the messages.

c. Machinery for automatically comparing letters in applying the coincidence test has been devised. Such machines greatly facilitate and speed up the procedure.

---

[5] At first thought the student might wonder why it is advisable or necessary to test message 3 against message 1 and message 2 separately before testing it against the combination of messages 1 and 2. The first two tests, it seems to him, might be omitted and time saved thereby. The matter will be explained in par. 43*f* (3).

43. Example of application of the κ test.—*a.* With the foregoing in mind, a practical example will now be given. The following messages, assumed to be the first 4 of a series of 30 messages, supposedly enciphered by a long keying sequence, but each message commencing at a different point in that sequence, are to be arranged so as to bring them into correct superimposition:

MESSAGE 1

```
P G L P N   H U F R K   S A U Q Q   A Q Y U O   Z A K G A   E O Q C N
P R K O V   H Y E I U   Y N B O N   N F D M W   Z L U K Q   A Q A H Z
M G C D S   L E A G C   J P I V J   W V A U D   B A H M I   H K O R M
L T F Y Z   L G S O G   K
```

MESSAGE 2

```
C W H P K   K X F L U   M K U R Y   X C O P H   W N J U W   K W I H L
O K Z T L   A W R D F   G D D E Z   D L B O T   F U Z N A   S R H H J
N G U Z K   P R C D K   Y O O B V   D D X C D   O G R G I   R M I C N
H S G G O   P Y A O Y   X
```

MESSAGE 3

```
W F W T D   N H T G M   R A A Z G   P J D S Q   A U P F R   O X J R O
H R Z W C   Z S R T E   E E V P X   O A T D Q   L D O Q Z   H A W N X
T H D X L   H Y I G K   V Y Z W X   B K O Q O   A Z Q N D   T N A L T
C N Y E H   T S C T
```

MESSAGE 4

```
T U L D H   N Q E Z Z   U T Y G D   U E D U P   S D L I O   L N N B O
N Y L Q Q   V Q G C D   U T U B Q   X S O S K   N O X U V   K C Y J X
C N J K S   A N G U I   F T O W O   M S N B Q   D B A I V   I K N W G
V S H I E   P
```

b. Superimposing[6] messages 1 and 2, beginning with their 1st letters,

```
                   5         10        15        20        25        30        35
No. 1.......  P G L P N H U F R K S A U Q Q A Q Y U O Z A K G A E O Q C N P R K O V
No. 2.......  C W H P K K X F L U M K U R Y X C O P H W N J U W K W I H L O K Z T L

                     40        45        50        55        60        65        70
No. 1.......  H Y E I U Y N B O N N F D M W Z L U K Q A Q A H Z M G C D S L E A G C
No. 2.......  A W R D F G D D E Z D L B O T F U Z N A S R H H J N G U Z K P R C D K

                     75        80        85        90        95        100
No. 1.......  J P I V J W V A U D B A H M I H K O R M L T F Y Z L G S O G K
No. 2.......  Y O O B V D D X C D O G R G I R M I C N H S G G O P Y A O Y X
```

the number of coincidences is found to be 8. Since the total number of comparisons is 101, the expected number, if the superimposition were correct, should be $101 \times .0667 = 6.7367$, or about 7 coincidences. The fact that the observed number of coincidences matches and is even greater than the expected number on the very first trial creates an element of suspicion: such good fortune is rarely the lot of the practical cryptanalyst. It is very unwise to stop at the first trial, *even if the results are favorable*, for this close agreement between theoretical and actual numbers

[6] The student will have to imagine the messages written out as continuous sequences on cross-section paper.

of coincidences might just be "one of those accidents." Therefore message 2 is shifted one space to the right, placing its 1st letter beneath the 2d letter of message 1. Again the number of coincidences is noted and this time it is found to be only 4. The total number of comparisons is now 100; the expected number is still about 7. Here the observed number of coincidences is considerably less than the expected number, and when the relatively small number of comparisons is borne in mind, the discrepancy between the theoretical and actual results is all the more striking. The hasty cryptanalyst might therefore jump to the conclusion that the 1st superimposition is actually the correct one. But only two trials have been made thus far and a few more are still advisable, for in this scheme of superimposing a series of messages it is absolutely essential that the very first superimpositions rest upon a perfectly sound foundation—otherwise subsequent work will be very difficult, if not entirely fruitless. Additional trials will therefore be made.

c. Message 2 is shifted one more space to the right and the number of coincidences is now found to be only 3. Once again message 2 is shifted, to the position shown below, and the observed number of coincidences jumps suddenly to 9.

```
                    5         10        15        20        25        30        35
No. 1------ P G L P N H U F R K S A U Q Q A Q Y U O Z A K G A E O Q C N P R K O V
No. 2------       C W H P K K X F L U M K U R Y X C O P H W N J U W K W I H L O K

                    40        45        50        55        60        65        70
No. 1------ H Y E I U Y N B O N N F D M W Z L U K Q A Q A H Z M G C D S L E A G C
No. 2------ Z T L A W R D F G D D E Z D L B O T F U Z N A S R H H J N G U Z K P R

                    75        80        85        90        95        100
No. 1------ J P I V J W V A U D B A H M I H K O R M L T F Y Z L G S O G K
No. 2------ C D K Y O O B V D D X C D O G R G I R M I C N H S G G O P Y A O Y X
```

The total number of comparisons is now 98, so that the expected number of coincidences is $98 \times .0667 = 6.5366$, or still about 7. The 2d and 3d superimpositions are definitely incorrect; as to the 1st and 4th, the latter gives almost 30 percent more coincidences than the former. Again considering the relatively small number of comparisons, this 30 percent difference in favor of the 4th superimposition as against the 1st is important. Further detailed explanation is unnecessary, and the student may now be told that it happens that the 4th superimposition is really correct; if the messages were longer, all doubt would be dispelled. The relatively large number of coincidences found at the 1st superimposition is purely accidental in this case.

d. The phenomenon noted above, wherein the observed number of coincidences shows a *sudden* increase in moving from an incorrect to a correct superimposition is not at all unusual, nor should it be unexpected, because there is only *one* correct superimposition, while *all other* superimpositions are entirely incorrect. In other words, a superimposition is either 100 percent correct or 100 percent wrong—and there are no gradations between these two extremes. Theoretically, therefore, the difference between the correct superimposition and any one of the many incorrect superimpositions should be very marked, since it follows from what has been noted above, that one cannot expect that the discrepancy between the actual and the theoretical number of coincidences should get smaller and smaller as one approaches closer and closer to the correct superimposition.[7] For if letters belonging to the same cipher alphabet are regarded

[7] The importance of this remark will be appreciated when the student comes to study longer examples, in which statistical expectations have a better opportunity to materialize.

as being members of the same family, so to speak, then the two letters forming the successive pairs of letters brought into superimposition by an incorrect placement of one message relative to another are total strangers to each other, brought together by pure chance. This happens time and again, as one message is slid against the other—until the correct superimposition is reached, whereupon in *every* case the two superimposed letters belong to the same family. There may be many different families (cipher alphabets) but the fact that in every case two members of the same family are present causes the marked jump in number of coincidences.

e. In shifting one message against another, the cryptanalyst may move to the right constantly, or he may move to the left constantly, or he may move alternately to the left and right from a selected initial point. Perhaps the latter is the best plan.

f. (1) Having properly superimposed messages 1 and 2, message 3 is next to be studied. Now it is of course possible to test the latter message against the combination of the former, without further ado. That is, ascertaining merely the total number of coincidences given by the superimposition of the 3 messages might be thought sufficient. But for reasons which will soon become apparent it is better, even though much more work is involved, first to test message 3 against message 1 alone and against message 2 alone. This will really not involve much additional work after all, since the two tests can be conducted simultaneously, because the proper superimposition of messages 1 and 2 is already known. If the tests against messages 1 and 2 separately at a given superimposition give good results, then message 3 can be tested, at that superimposition, against messages 1 and 2 combined. That is, all 3 messages are tested as a single set. Since, according to the scheme outlined, a set of three closely related tests is involved, one might as well systematize the work so as to save time and effort, if possible. With this in view a diagram such as that shown in Figure 11*a* is made and in it the coincidences are recorded in the appropriate cells, to show separately the coincidences between messages 1 and 2, 1 and 3, 2 and 3, for each superimposition tested. The number of tallies in the cell 1–2 is the same at the beginning of all the tests; it has already been found to be 9. Therefore, 9 tallies are inserted in cell 1–2 to begin with. A column which shows identical letters in messages 1 and 3 yields a single tally for cell 1–3; a column which shows identical letters in messages 2 and 3 yields a single tally for cell 2–3. Only when a superimposition yields 3 identical letters in a column, is a tally to be recorded simultaneously in cells 1–3 and 2–3, since the presence of 3 identical letters in the column yields 3 coincidences.

	1	2	3
1	×	~~////~~ ////	///
2	×	×	///
3	×	×	×

FIGURE 11*a*.

(2). Let message 3 be placed beneath messages 1 and 2 combined, so that the 1st letter of message 3 falls under the 1st letter of message 1. (It is advisable to fasten the latter in place so that they cannot easily be disturbed.) Thus:

```
              1  2  3  4  5  6  7  8  9 10 11 12 13 14 15 16 17 18 19 20 21 22 23 24 25 26 27
1..........   P  G  L  P  N  H  U  F  R  K  S  A  U  Q  Q  A  Q  Y  U  O  Z  A  K  G  A  E  O
2..........            C  W  H  P  K  K  X  F  L  U  M  K  U  R  Y  X  C  O  P  H  W  N  J  U
3..........   W  F  W  T  D  N  H  T  G  M  R  A  A  Z  G  P  J  D  S  Q  A  U  P  F  R  O  X

             28 29 30 31 32 33 34 35 36 37 38 39 40 41 42 43 44 45 46 47 48 49 50 51 52 53 54
1..........   Q  C  N  P  R  K  O  V  H  Y  E  I  U  Y  N  B  O  N  N  F  D  M  W  Z  L  U  K
2..........   W  K  W  I  H  L  O  K  Z  T  L  A  W  R  D  F  G  D  D  E  Z  D  L  V  O  T  F
3..........   J  R  O  H  R  Z  W  C  Z  S  R  T  E  E  E  V  P  X  O  A  T  D  Q  L  D  O  Q

             55 56 57 58 59 60 61 62 63 64 65 66 67 68 69 70 71 72 73 74 75 76 77 78 79 80 81
1..........   Q  A  Q  A  H  Z  M  G  C  D  S  L  E  A  G  C  J  P  I  V  J  W  V  A  U  D  B
2..........   U  Z  N  A  S  R  H  H  J  N  G  U  Z  K  P  R  C  D  K  Y  O  O  B  V  D  D  X
3..........   Z  H  A  W  N  X  T  H  D  X  L  H  Y  I  G  K  V  Y  Z  W  X  B  K  O  Q  O  A

             82 83 84 85 86 87 88 89 90 91 92 93 94 95 96 97 98 99 100 101 102 103 104
1..........   A  H  M  I  H  K  O  R  M  L  T  F  Y  Z  L  G  S  O  G   K
2..........   C  D  O  G  R  G  I  R  M  I  C  N  H  S  G  G  O  P  Y   A   O   Y   X
3..........   Z  Q  N  D  T  N  A  L  T  C  N  Y  E  H  T  S  C  T
```

	1	2	3
1	×	卌 ////	///
2	×	×	///
3	×	×	×

FIGURE 11*b*.

The successive columns are now examined and the coincidences are recorded, remembering that only coincidences between messages 1 and 3, and between messages 2 and 3 are now to be tabulated in the diagram. The results for this first test are shown in Figure 11*b*. This superimposition yields but 3 coincidences between messages 1 and 3, and the same number between messages 2 and 3. The total numbers of comparisons are then noted and the following table is drawn up:

Combination	Total number of comparisons	Number of coincidences		Discrepancy
		Expected	Observed	
				Percent
Messages 1 and 3	99	About 7	3	−57
Messages 2 and 3	96	About 6	3	−50
Messages 1, 2, and 3	293	About 20	15	−21

(3) The reason for the separate tabulation of coincidences between messages 1 and 3, 2 and 3, and 1, 2, and 3 should now be apparent. Whereas the observed number of coincidences is 57 percent below the expected number of coincidences in the case of messages 1 and 3 alone, and 50 percent below in the case of messages 2 and 3 alone, the discrepancy between the expected and observed numbers is not quite so marked (—21 percent) when all three messages are considered together, because the relatively high number of coincidences between messages 1 and 2, which are correctly superimposed, serves to counterbalance the low numbers of coincidences between 1 and 3, and 2 and 3. *Thus, a correct superimposition for one of the three combinations may yield such good results as to mask the bad results for the other two combinations.*

(4) Message 3 is then shifted one space to the right, and the same procedure is followed as before. The results are shown below:

```
                  5         10        15        20        25        30        35
No. 1....... P G L P N H U F R K S A U Q Q A Q Y U O Z A K G A E O Q C N P R K O V
No. 2.......       C W H P K K X F L U M K U R Y X C O P H W N J U W K W I H L O K
No. 3.......   W F W T D N H T G M R A A Z G P J D S Q A U P F R O X J R O H R Z W

                  40        45        50        55        60        65        70
No. 1....... H Y E I U Y N B O N N F D M W Z L U K Q A Q A H Z M G C D S L E A G C
No. 2....... Z T L A W R D F G D D E Z D L B O T F U Z N A S R H H J N G U Z K P R
No. 3....... C Z S R T E E E V P X O A T D Q L D O Q Z H A W N X T H D X L H Y I G

                  75        80        85        90        95        101
No. 1....... J P I V J W V A U D B A H M I H K O R M L T F Y Z L G S O G K
No. 2....... C D K Y O O B V D D X C D O G R G I R M I C N H S G G O P Y A O Y X
No. 3....... K V Y Z W X B K O Q O A Z Q N D T N A L T C N Y E H T S C T
```

	1	2	3
1	×	𝍸 ////	𝍸 𝍸
2	×	×	𝍸 /
3	×	×	×

FIGURE 11c.

Combination	Total number of comparisons	Number of coincidences		Discrepancy
		Expected	Observed	
				Percent
Messages 1 and 3	99	About 7	10	+43
Messages 2 and 3	97	About 6	6	0
Messages 1, 2, and 3	294	About 20	25	+25

Note how well the observed and expected numbers of coincidences agree in all three combinations. Indeed, the results of this test are so good that the cryptanalyst might well hesitate to make any more tests.

(5) Having ascertained the relative positions of 3 messages, the fourth message is now studied. Here are the results for the correct superimposition.

```
                     5         10        15        20        25        30        35
No. 1....... P G L P N H U F R K S A U Q Q A Q Y U O Z A K G A E O Q C N P R K O V
No. 2.......       C W H P K K X F L U M K U R Y X C O P H W N J U W K W I H L O K
No. 3.......   W F W T D N H T G M R A A Z G P J D S Q A U P F R O X J R O H R Z W
No. 4.......     T U L D H N Q E Z Z U T Y G D U E D U P S D L I O L N N B O N Y L

                     40        45        50        55        60        65        70
No. 1....... H Y E I U Y N B O N N F D M W Z L U K Q A Q A H Z M G C D S L E A G C
No. 2....... Z T L A W R D F G D D E Z D L B O T F U Z N A S R H H J N G U Z K P R
No. 3....... C Z S R T E E E V P X O A T D Q L D O Q Z H A W N X T H D X L H Y I G
No. 4....... Q Q V Q G C D U T U B Q X S O S K N O X U V K C Y J X C N J K S A N G

                     75        80        85        90        95          101
No. 1....... J P I V J W V A U D B A H M I H K O R M L T F Y Z L G S O G K
No. 2....... C D K Y O O B V D D X C D O G R G I R M I C N H S G G O P Y A O Y X
No. 3....... K V Y Z W X B K O Q O A Z Q N D T N A L T C N Y E H T S C T
No. 4....... U I F T O W O M S N B Q D B A I V I K N W G V S H I E P
```

	1	2	3	4
1	×	𝍸 ||||	𝍸 𝍸	𝍸 ||
2	×	×	𝍸 |	𝍸 |
3	×	×	×	𝍸
4	×	×	×	×

FIGURE 11*d*

Combination	Total number of comparisons	Number of coincidences		Discrepancy
		Expected	Observed	
				Percent
Messages 1 and 4	96	About 6	7	+16
Messages 2 and 4	95	About 6	7	+16
Messages 3 and 4	96	About 6	5	−16
Messages 1, 2, 3, and 4	581	About 39	44	+10

The results for an incorrect superimposition (1st letter of message 4 under 4th letter of message 1) are also shown for comparison:

```
                 5          10            15            20            25            30          33
No. 1....... P G L P N H U F R K S A U Q Q A Q Y U O Z A K G A E O Q C N P R K O V
No. 2.......         C W H P K K X F L U M K U R Y X C O P H W N J U W K W I H L O K
No. 3.......   W F W T D N H T G M R A A Z G P J D S Q A U P F R O X J R O H R Z W
No. 4.......         T U L D H N Q E Z Z U T Y G D U E D U P S D L I O L N N B O N Y

                 40          45          50          55            60            65          70
No. 1....... H Y E I U Y N B O N N F D M W Z L U K Q A Q A H Z M G C D S L E A G C
No. 2....... Z T L A W R D F G D D E Z D L B O T F U Z N A S R H H J N G U Z K P R
No. 3....... C Z S R T E E E V P X O A T D Q L D O Q Z H A W N X T H D X L H Y I G
No. 4....... L Q Q V Q G C D U T U B Q X S O S K N O X U V K C Y J X C N J K S A N

                 75          80          85          90          95          101
No. 1....... J P I V J W V A U D B A H M I H K O R M L T F Y Z L G S O G K
No. 2....... C D K Y O O B V D D X C D O G R G I R M I C N H S G G O P Y A O Y X
No. 3....... K V Y Z W X B K O Q O A Z Q N D T N A L T C N Y E H T S C T
No. 4....... G U I F T O W O M S N B Q D B A I V I K N W G V S H I E P
```

	1	2	3	4
1	×	卌 ////	卌 卌	///
2	×	×	卌 /	///
3	×	×	×	/
4	×	×	×	×

FIGURE 11e

Combination	Total number of comparisons	Number of coincidences		Discrepancy
		Expected	Observed	
				Percent
Messages 1 and 4	96	About 6	3	−50
Messages 2 and 4	96	About 6	3	−50
Messages 3 and 4	96	About 6	2	−83
Messages 1, 2, 3 and 4	582	About 39	33	−18

(6) It is believed that the procedure has been explained with sufficient detail to make further examples unnecessary. The student should bear in mind always that as he adds messages to the superimposition diagram it is necessary that he recalculate the number of comparisons so that the correct expected or theoretical number of coincidences will be before him to compare with the observed number. In adding messages he should see that the results of the separate tests are consistent, as well as those for the combined tests, otherwise he may be led astray at times by the overbalancing effect of the large number of coincidences for the already ascertained, correct superimpositions.

44. Subsequent steps.—*a*. In paragraph 43*a* four messages were given of a series supposedly enciphered by a long keying sequence, and the succeeding paragraphs were devoted to an explanation of the preparatory steps in the solution. The messages have now been properly superimposed, so that the text has been reduced to monoalphabetic columnar form, and the matter is now to be pursued to its ultimate stages.

b. The four messages employed in the demonstration of the principles of the κ test have served their purpose. The information that they are messages enciphered by an intelligible running key, by reversed standard cipher alphabets, was withheld from the student, for pedagogical reasons. Were the key a random sequence of letters instead of intelligible text, the explanation of the coincidence test would have been unchanged in the slightest particular, so far as concerns the mechanics of the text itself. Were the cipher alphabets unknown, mixed alphabets, the explanation of the κ test would also have been unchanged in the slightest particular. But, as stated before, the four messages actually represent encipherments by means of an intelligible running key, by reversed standard alphabets; they will now be used to illustrate the solution of cases of this sort.

c. Assuming now that the cryptanalyst is fully aware that the enemy is using the running-key system with reversed standard alphabets (obsolete U. S. Army cipher disk), the method of solution outlined in paragraph 38 will be illustrated, employing the first of the four messages referred to above, that beginning PGLPN HUFRK SAUQQ. The word DIVISION will be taken as a probable word and tested against the key, beginning with the very first letter of the message. Thus:

```
Cipher text..................... P G L P N H U F R K S A U Q Q . .
Assumed plain text.............. D I V I S I O N . . . . . . . . .
Resultant key text.............. S O G X F . . . . . . . . . . . .
```

The resultant key text is unintelligible and the word DIVISION is shifted one letter to the right.

```
Cipher text..................... P G L P N H U F R K S A U Q Q . .
Assumed plain text.............. . D I V I S I O N . . . . . . . .
Resultant key text.............. . J T K . . . . . . . . . . . . .
```

Again the resultant key text is unintelligible and the hypothetical word DIVISION is shifted once more. Continuation of this process to the end of the message proves that the word is not present. Another probable word is assumed: REGIMENT. When the point shown below is reached, note the results:

```
Cipher text..................... P G L P N H U F R K S A U Q Q . .
Assumed plain text.............. . . . . R E G I M E N T . . . . .
Resultant key text.............. . . . . E L A N D O F T . . . . .
```

It certainly looks as though intelligible text were being obtained as key text. The words LAND OF T . . . suggest that THE be tried. The key letters HE give NO, making the plain text read . . . . REGIMENT NO . . . . The four spaces preceding REGIMENT suggest such words as HAVE, SEND, MOVE, THIS, etc. A clue may be found by assuming that the E before LAND in the key is part of the word THE. Testing it on the cipher text gives IS for the plain text, which certainly indicates that the message begins with the word THIS. The latter yields IN for the first two key letters. And so on, the process of checking one text against the other continuing until the entire message and the key text have been reconstructed.

d. Thus far the demonstration has employed but one of the four messages available for solution. When the reconstruction process is applied to all four simultaneously it naturally goes much faster, with reduced necessity for assuming words after an initial entering wedge has been driven into one message. For example, note what happens in this case just as soon as the word REGIMENT is tried in the proper place:

	Key text	.	.	.	.	E	L	A	N	D	O	F	T	.	.	.	. . .
No. 1	Cipher text	P	G	L	P	N	H	U	F	R	K	S	A	U	Q	Q	. . .
	Plain text	.	.	.	.	R	E	G	I	M	E	N	T	.	.	.	. . .
No. 2	Cipher text	.	.	.	C	W	H	P	K	K	X	F	L	U	M	K	. . .
	Plain text	.	.	.	.	I	E	L	D	T	R	A	I	.	.	.	. . .
No. 3	Cipher text		W	F	W	T	D	N	H	T	G	M	R	A	A	Z	. . .
	Plain text		.	.	.	L	I	N	G	K	I	T	C	.	.	.	. . .
No. 4	Cipher text			T	U	L	D	H	N	Q	E	Z	Z	U	T	Y	. . .
	Plain text			.	.	T	I	T	A	N	K	G	U	.	.	.	. . .

It is obvious that No. 2 begins with FIELD TRAIN; No. 3, with ROLLING KITCHEN; No. 4 with ANTITANK GUN. These words yield additional key letters, the latter suggest additional plain text, and thus the process goes on until the solution is completed.

e. But now suppose that the key text that has been actually employed in encipherment is not intelligible text. The process is still somewhat the same, only in this case one must have at least two messages in the same key. For instead of checking a hypothetical word (assumed to be present in one message) against the key, *the same kind of a check is made against the other message or messages*. Assume, for instance, that in the case just described the key text, instead of being intelligible text, were a series of letters produced by applying a rather complex transposition to an originally intelligible key text. Then if the word REGIMENT were assumed to be present in the proper place in message No. 1 the resultant key letters would yield an unintelligible sequence. But these key letters when applied to message No. 2 would nevertheless yield IELDTRAI; when applied to message No. 3, LINGKITC, and so on. In short, the *text of one message is checked against the text of another message or messages;* if the originally assumed word is correct, then plain text will be found in the other messages.[8]

---

[8] Perhaps this is as good a place as any to make some observations which are of general interest in connection with the running-key principle, and which have no doubt been the subject of speculation on the part of some students. Suppose a basic, unintelligible, random sequence of keying characters which is not derived from the interaction of two or more shorter keys and which *never repeats* is employed *but once* as a key for encipherment. Can a cryptogram enciphered in such a system be solved? The answer to this question must unqualifiedly be this: even if the cipher alphabets are known sequences, cryptanalytic science is certainly powerless to attack such a cryptogram. Furthermore, so far as can now be discerned, no method of attack is likely ever to be devised. Short of methods based upon the alleged phenomena of telepathy—the very objective existence of which is denied by most "sane" investigators today—it is impossible for the present author to conceive of any way of attacking such a cryptogram.

This is a case (and perhaps the only case) in which the impossibility of cryptanalysis is mathematically demonstrable. Two things are involved in a complete solution in mathematics: not only must a satisfactory (logical) answer to the problem be offered, but also it must be demonstrated that the answer offered is *unique*, that is, the only possible one. (The mistake is often made that the latter phase of what constitutes a valid solution is overlooked—and this is the basic error which numerous alleged Bacon-Shakespeare "cryptographers" commit.) To attempt to solve a cryptogram enciphered in the manner indicated is analogous to an attempt to find a unique solution for a single equation containing two unknowns, with absolutely no data available for solution other than those given by that equation itself. It is obvious that no unique solution is possible in such a case, since *any one quantity whatsoever* may be chosen for one of the unknowns and the other will follow as a consequence. Therefore an infinite number of different answers, all equally valid, is possible. In the case of a *f.* All the foregoing work is, of course, based upon a knowledge of the cipher alphabets employed in the encipherment. What if the latter are unknown sequences? It may be stated at once that not much could be done with but four messages, even after they had been superimposed correctly, for the most that one would have in the way of data for the solution of the individual columns of text would be four letters per alphabet—which is not nearly enough. Data for solution by indirect symmetry by the detection of isomorphs cannot be expected, for no isomorphs are produced in this system. Solution can be reached only if there is sufficient text to permit of the analysis of the columns of the superimposition diagram. When there is this amount of text there are also repetitions which afford bases for the assumption of probable words. Only then, and after the values of a few cipher letters have been established can indirect symmetry be applied to facilitate the reconstruction of the primary components—if used.

g. Even when the volume of text is great enough so that each column contains say 15 to 20 letters, the problem is still not an easy one. But frequency distributions with 15 to 20 letters can usually be studied statistically, so that if two distributions present similar characteristics, the latter may be used as a basis for combining distributions which pertain to the same cipher alphabet. The next section will be devoted to a detailed treatment of the implications of the last statement.

---

cryptogram enciphered in the manner indicated, there is the equivalent of an equation with two unknowns; the key is one of the unknowns, the plain text is the other. One may conjure up an infinite number of different plain texts and offer any one of them as a "solution." One may even perform the perfectly meaningless labor of reconstructing the "key" for this selected "solution"; but since there is no way of proving from the cryptogram itself, or from the reconstructed key (which is unintelligible) whether the "solution" so selected is *the* actual plain text, all of the infinite number of "solutions" are equally valid. Now since it is inherent in the very idea of cryptography as a practical art that there must and can be only *one* actual solution (or plain text), and since none of this infinite number of different solutions can be proved to be *the one and only* correct solution, therefore, our common sense rejects them one and all, and it may be said that a cryptogram enciphered in the manner indicated is absolutely impossible to solve.

It is perhaps unnecessary to point out that the foregoing statement is no longer true when the running key constitutes intelligible text, or if it is used to encipher more than one message, or if it is the secondary resultant of the interaction of two or more short primary keys which go through cycles themselves. For in these cases there is additional information available for the delimitation of one of the pair of unknowns, and hence a unique solution becomes possible.

Now although the running-key system described in the first paragraph represents the ultimate goal of cryptographic security and is the ideal toward which cryptographic experts have striven for a long time, there is a wide abyss to be bridged between the recognition of a theoretically perfect system and its establishment as a practical means of secret intercommunication. For the mere mechanical details involved in the production, reproduction, and distribution of such keys present difficulties which are so formidable as to destroy the effectiveness of the method as a system of secret intercommunication suitable for groups of correspondents engaged in a voluminous exchange of messages.

SECTION XII

THE "CROSS-PRODUCT SUM" OR "χ TEST"[1]

45. Preliminary remarks.—*a.* The real purpose of making the coincidence test in cases such as that studied in the preceding section is to permit the cryptanalyst to arrange his data so as to circumvent the obstacle which the enemy, by adopting a complicated polyalphabetic scheme of encipherment, places in the way of solution. The essence of the matter is that by dealing individually with the respective columns of the superimposition diagram the cryptanalyst has arranged the polyalphabetic text so that it can be handled as though it were monoalphabetic. Usually, the solution of the latter is a relatively easy matter, especially if there is sufficient text in the columns, or if the letters within certain columns can be combined into single frequency distributions, or if some cryptographic relationship can be established between the columns.

b. It is obvious that merely ascertaining the correct relative positions of the separate messages of a series of messages in a superimposition diagram is only a means to an end, and not an end in itself. The purpose is, as already stated, to reduce the complex, heterogeneous, polyalphabetic text to simple, homogeneous, monoalphabetic text. But the latter can be solved only when there are sufficient data for the purpose—and that depends often upon the type of cipher alphabets involved. The latter may be the secondary alphabets resulting from the sliding of the normal sequence against its reverse, or a mixed component against the normal, and so on. The student has enough information concerning the various cryptanalytic procedures which may be applied, depending upon the circumstances, in reconstructing different types of primary components and no more need be said on this score at this point.

c. The student should, however, realize one point which has thus far not been brought specifically to his attention. Although the superimposition diagram referred to in the preceding subparagraph may be composed of many columns, there is often only a relatively small number of *different* cipher alphabets involved. For example, in the case of two primary components of 26 letters each there is a maximum of 26 secondary cipher alphabets. Consequently, it follows that in such a case if a superimposition diagram is composed of say 100 columns, certain of those columns must represent similar secondary alphabets. There may, and probably will be, no regularity of recurrence of these repeated secondaries, for they are used in a manner directly governed by the letters composing the words of the key text or the elements composing the keying sequence.

d. But the latter statement offers an excellent clue. It is clear that the number of times a given secondary alphabet is employed in such a superimposition diagram depends upon the com-

[1] The χ test, presented in this section, as well as the Φ test, presented in Section XIV, were first described in an important paper, *Statistical Methods in Cryptanalysis*, 1935, by Solomon Kullback, Ph. D., Associate Cryptanalyst, Signal Intelligence Service. I take pleasure in acknowledging my indebtedness to Dr. Kullback's paper for the basic material used in my own exposition of these tests, as well as for his helpful criticisms thereof while in manuscript.

position of the key text. Since in the case of a running-key system using a book as a key the key text constitutes intelligible text, it follows that *the various secondary alphabets will be employed with frequencies which are directly related to the respective frequencies of occurrence of letters in normal plain text.* Thus, the alphabet corresponding to key letter E should be the most frequently used; the alphabet corresponding to key letter T should be next in frequency, and so on. From this it follows that instead of being confronted with a problem involving as many different secondary cipher alphabets as there are columns in the superimposition diagram, the cryptanalyst will usually have not over 26 such alphabets to deal with; and allowing for the extremely improbable repetitive use of alphabets corresponding to key letters J, K, Q, X, and Z, it is likely that the cryptanalyst will have to handle only about 19 or 20 secondary alphabets.

e. Moreover, since the E secondary alphabet will be used most frequently and so on, it is possible for the cryptanalyst to study the various distributions for the columns of the superimposition diagram with a view to assembling those distributions which belong to the same cipher alphabet, thus making the actual determination of values much easier in the combined distributions than would otherwise be the case.

f. However, if the keying sequence does not itself constitute intelligible text, even if it is a random sequence, the case is by no means hopeless of solution—provided there is sufficient text within columns so that the columnar frequency distributions may afford indications enabling the cryptanalyst to amalgamate a large number of small distributions into a smaller number of larger distributions.

g. In this process of assembling or combining individual frequency distributions which belong to the same cipher alphabet, recourse may be had to a procedure merely alluded to in connection with previous problems, and designated as that of "matching" distributions. The next few paragraphs will deal with this important subject.

46. The nature of the "Cross-product sum" or "χ (Chi) test" in cryptanalysis.—*a:* The student has already been confronted with cases in which it was necessary or desirable to reduce a large number of frequency distributions to a smaller number by identifying and amalgamating distributions which belong to the same cipher alphabet. Thus, for example, in a case in which there are, say, 15 distributions but only, say, 5 separate cipher alphabets, the difficulty in solving a message can be reduced to a considerable degree provided that of the 15 distributions those which belong together can be identified and allocated to the respective cipher alphabets to which they apply.

b. This process of identifying distributions which belong to the same cipher alphabet involves a careful examination and comparison of the various members of the entire set of distributions to ascertain which of them present sufficiently similar characteristics to warrant their being combined into a single distribution applicable to one of the cipher alphabets involved in the problem. Now when the individual distributions are fairly large, say containing over 50 or 60 letters, the matter is relatively easy for the experienced cryptanalyst and can be made by the eye; but when the distributions are small, each containing a rather small number of letters, ocular comparison and identification of two or more distributions as belonging to the same alphabet become quite difficult and often inconclusive. In any event, the time required for the successful reduction of a multiplicity of individual small distributions to a few larger distributions is, in such cases, a very material factor in determining whether the solution will be accomplished in time to be of actual value or merely of historical interest.

c. However, a certain statistical test, called the "cross-product sum" or "χ test", has been devised, which can be brought to bear upon this question and, by methods of mathematical comparison, eliminate to a large degree the uncertainties of the ocular method of matching and combining frequency distributions, thus in many cases materially reducing the time required for solution of a complex problem.

d. It is advisable to point out, however, that the student must not expect too much of a mathematical method of comparing distributions, because there are limits to the size of distributions to be matched below which these methods will not be effective. If two distributions contain some similar characteristics the mathematical method will merely afford a quantitative measure of the degree of similarity. Two distributions may actually pertain to the same cipher alphabet but, as occasionally happens, they may not present any external evidences of this relationship, in which case *no* mathematical method can indicate the fact that the two distributions are really similar and belong to the same alphabet.

47. Derivation of the χ test.—*a*. Consider the following plain-text distribution of 50 letters:

A B C D E F G H I J K L M N O P Q R S T U V W X Y Z

In a previous text [2] it was shown that the chance of drawing two identical letters in normal English telegraphic plain text is the sum of the squares of the relative probabilities of occurrence of the 26 letters in such text, which is .0667. That is, the probability of monographic coincidence in English telegraphic plain text is $\kappa_p = .0667$. In the message to which the foregoing distribution of 50 letters applies, the number of possible pairings (comparisons) that can be made between single letters is $\frac{50 \times 49}{2} = 1{,}225$. According to the theory of coincidences there should, therefore be $1{,}225 \times .0667 = 81.7065$ or approximately 82 coincidences of single letters. Examining the distribution it is found that there are 83 coincidences, as shown below:

A B C D E F G H I J K L M N O P Q R S T U V W X Y Z
3+0+0+1+21+0+0+1+3+0+0+0+1+10+15+0+0+1+10+15+1+0+1+0+0+0=83

The actual number of coincidences agrees very closely with the theoretical number, which is of course to be expected, since the text to which the distribution applied has been indicated as being normal plain text.

b. In the foregoing simple demonstration, let the number of comparisons that can be made in the distribution be indicated symbolically by $\frac{N(N-1)}{2}$, where N=the total number of letters in the distribution. Then the expected number of coincidences may be written as $\frac{.0067N(N-1)}{2}$, which may then be rewritten as

$$\text{(I)} \qquad \frac{.0067N^2 - .0667N}{2}.$$

c. Likewise, if f_A represents the number of occurrences of A in the foregoing distribution, then the number of coincidences for the letter A may be indicated symbolically by $\frac{f_A(f_A-1)}{2}$. And similarly, the number of coincidences for the letter B may be indicated by $\frac{f_B(f_B-1)}{2}$, and so on down to $\frac{f_Z(f_Z-1)}{2}$. The total number of actual coincidences found in the distribution is, of course, the sum of $\frac{f_A(f_A-1)}{2} + \frac{f_B(f_B-1)}{2} + \ldots \frac{f_Z(f_Z-1)}{2}$. If the symbol f_θ is used to indicate any of the letters A, B, . . . Z, and the symbol Σ is used to indicate that the sum of all the

[2] *Military Cryptanalysis, Part II*, Appendix 2.

elements that follow this sign is to be found, then the sum of the actual coincidences noted in the distribution may be indicated thus: $\sum\frac{f_\theta(f_\theta-1)}{2}$, which may be rewritten as

(II) $$\sum\frac{f_\theta^2-f_\theta}{2}$$

d. Now although derived from different sources, the two expressions labeled (I) and (II) above are equal, or should be equal, in normal plain text. Therefore, one may write:

$$\sum\frac{f_\theta^2-f_\theta}{2}=\frac{.0667N^2-.0667N}{2}$$

Simplifying this equation:

(III) $$\Sigma f_\theta^2-\Sigma f_\theta=.0667N^2-.0667N$$

e. Now $\Sigma f_\theta=N$.

Therefore, expression (III) may be written as

(IV) $$\Sigma f_\theta^2-N=.0667N^2-.0667N,$$

which on reduction becomes:

(V) $$\Sigma f_\theta^2=.0667N^2+.9333N$$

This equation may be read as "the sum of the squares of the absolute frequencies of a distribution is equal to .0667 times the square of the total number of letters in the distribution, plus .9333 times the total number of letters in the distribution." The letter S_2 is often used to replace the symbol Σf_θ^2.

f. Suppose two monoalphabetic distributions are thought to pertain to the same cipher alphabet. Now if they actually do belong to the same alphabet, and if they are correctly[3] combined into a single distribution, the latter must still be monoalphabetic in character. That is, again representing the individual letter frequencies in one of these distributions by the general symbol f_{θ_1} the individual letter frequencies in the other distribution by f_{θ_2}, and the total frequency in the first distribution by N_1, that in the second distribution by N_2, then

(VI) $$\Sigma(f_{\theta_1}+f_{\theta_2})^2=.0667(N_1+N_2)^2+.9333(N_1+N_2)$$

Expanding the terms of this equation:

(VII) $$\Sigma f_{\theta_1}^2+2\Sigma f_{\theta_1}f_{\theta_2}+\Sigma f_{\theta_2}^2=.0667(N_1^2+2N_1N_2+N_2^2)+.9333N_1+.9333N_2$$

But from equation (V):

$$\Sigma f_{\theta_1}^2=.0667N_1^2+.9333N_1 \text{ and}$$

$$\Sigma f_{\theta_2}^2=.0667N_2^2+.9333N_2,$$

so that equation (VII) may be rewritten thus:

$$.0667N_1^2+.9333N_1+2\Sigma f_{\theta_1}f_{\theta_2}+.0667N_2^2+.9333N_2=$$
$$.0667(N_1^2+2N_1N_2+N_2^2)+.9333N_1+.9333N_2$$

[3] By "correctly" is meant that the two distributions are slid relative to each other to their proper superimposition.

Reducing to simplest terms by cancelling out similar expressions:

$$2\Sigma f_{\theta_1} f_{\theta_2} = .0667(2N_1N_2), \text{ or}$$

(VIII)
$$\frac{\Sigma f_{\theta_1} f_{\theta_2}}{N_1N_2} = .0667$$

g. The last equation thus permits of establishing an expected value for the sum of the products of the corresponding frequencies of the two distributions being considered for amalgamation. The cross-product sum or χ *test for matching two distributions is based upon equation (VIII).*

48. Applying the χ **test in matching distributions.**—*a.* Suppose the following two distributions are to be matched:

f_1.............. A B C D E F G H I J K L M N O P Q R S T U V W X Y Z

f_2.............. A B C D E F G H I J K L M N O P Q R S T U V W X Y Z

Let the frequencies be juxtaposed, for convenience in finding the sum of the cross products. Thus:

	A	B	C	D	E	F	G	H	I	J	K	L	M	N	O	P	Q	R	S	T	U	V	W	X	Y	Z	
f_{θ_1}	1	4	0	3	0	1	0	0	1	0	0	1	0	0	1	0	0	3	2	2	1	0	1	3	0	2	$N_1=26$
f_{θ_2}	0	2	0	0	0	3	0	0	1	0	1	0	0	1	1	0	0	3	1	1	0	0	0	0	1	2	$N_2=17$
$f_{\theta_1} f_{\theta_2}$	0	8	0	0	0	3	0	0	1	0	0	0	0	0	1	0	0	9	2	2	0	0	0	0	0	4	

In this case $\Sigma f_{\theta_1} f_{\theta_2} = 8+3+1+1+9+2+2+4=30$

$$N_1N_2 = 26 \times 17 = 442$$

$$\frac{\Sigma f_{\theta_1} f_{\theta_2}}{N_1N_2} = \frac{30}{422} = .0711$$

b. The fact that the quotient (.0711) agrees very closely with the expected value (.0667) means that the two distributions very probably belong together or are properly matched. Note the qualifying phrase "very probably." It implies that there is no certainty about this business of matching distributions by mathematical methods. The mathematics serve only as measuring devices, so to speak, which can be employed to measure the degree of similarity that exists.

c. Instead of dividing $\Sigma f_{\theta_1} f_{\theta_2}$ by N_1N_2 and seeing how closely the quotient approximates the value .0667 or .0385, one may set up an expected value for $\Sigma f_{\theta_1} f_{\theta_2}$ and compare it with the observed value. Thus, in the foregoing example .0667 (N_1N_2)=.0667×422=28.15; the observed value of $\Sigma f_{\theta_1} f_{\theta_2}$ is 30 and therefore the agreement between the expected and the observed values is quite close, indicating that the two distributions are probably properly matched.

d. There are other mathematical or statistical tests for matching, in addition to the χ test. Moreover, it is possible to go further with the χ test and find a measure of the reliance that may be placed upon the value obtained; but these points will be left for future discussion in subsequent texts.

462861 O - 42 - 6

e. One more point will, however, here be added in connection with the χ test. Suppose the very same two distributions in subparagraph *a* are again juxtaposed, but with f_{θ_2} shifted one interval to the left of the position shown in the subparagraph of reference. Thus:

f_{θ_1}	1	4	0	3	0	1	0	0	1	0	0	1	0	0	1	0	0	3	2	2	1	0	1	3	0	2	$N_1=26$
	A	B	C	D	E	F	G	H	I	J	K	L	M	N	O	P	Q	R	S	T	U	V	W	X	Y	Z	
f_{θ_2}	B	C	D	E	F	G	H	I	J	K	L	M	N	O	P	Q	R	S	T	U	V	W	X	Y	Z	A	$N_2=17$
	2	0	0	0	3	0	0	1	0	1	0	0	1	1	0	0	3	1	1	0	0	0	0	1	2	0	

Here $\Sigma f_{\theta_1} f_{\theta_2}=2+3+2+3=10$ and $\frac{\Sigma f_{\theta_1} f_{\theta_2}}{N_1 N_2}=\frac{10}{442}=.0226$

The observed ratio (.0226) is so much smaller than the expected (.0667) that it can be said that if the two distributions pertain to the same primary components they are not properly superimposed. *In other words, the χ test may also be applied in cases where two or more frequency distributions must be shifted relatively in order to find their correct superimposition.* The theory underlying this application of the χ test is, of course, the same as before: two monoalphabetic distributions when properly combined will yield a single distribution which should still be monoalphabetic in character. In applying the χ test in such cases it may be necessary to shift two 26-element distributions to various superimpositions, make the χ test for each superimposition, and take as correct that one which yields the best value for the test.

f. The nature of the problem will, of course, determine whether the frequency distributions which are to be matched should be compared (1) by direct superimposition, that is, setting the A to Z tallies of one distribution directly opposite the corresponding tallies of the other distribution, as in subparagraph *a*, or (2) by shifted superimposition, that is, keeping the A to Z tallies of the first distribution fixed and sliding the whole sequence of tallies of the second distribution to various superimpositions against the first.

SECTION XIII

APPLYING THE CROSS-PRODUCT SUM OR χ TEST

49. Study of a situation in which the χ test may be applied.—*a*. A simple demonstration of how the χ-test is applied in matching frequency distributions may now be set before the student. The problem involved is the solution of cryptograms enciphered according to the progressive-alphabet system (par. 36*b*), with secondary alphabets derived from the interaction of two identical mixed primary components. It will be assumed that the enemy has been using a system of this kind and that the primary components are changed daily.

b. Before attacking an actual problem of this type, suppose a few minutes be devoted to a general analysis of its elements. It is here assumed that the primary components are based upon the HYDRAULIC . . . Z sequence and that the cipher component is shifted toward the right one step at a time. Consider a cipher square such as that shown in Figure 12, which is applicable to the type of problem under study. It has been arranged in the form of a deciphering square. In this square, *the horizontal sequences are all identical but merely shifted relatively; the letters inside the square are plain-text letters.*

ALPHABET NO.

Cipher Letter	1	2	3	4	5	6	7	8	9	10	11	12	13	14	15	16	17	18	19	20	21	22	23	24	25	26
A	A	U	L	I	C	B	E	F	G	J	K	M	N	O	P	Q	S	T	V	W	X	Z	H	Y	D	R
B	B	E	F	G	J	K	M	N	O	P	Q	S	T	V	W	X	Z	H	Y	D	R	A	U	L	I	C
C	C	B	E	F	G	J	K	M	N	O	P	Q	S	T	V	W	X	Z	H	Y	D	R	A	U	L	I
D	D	R	A	U	L	I	C	B	E	F	G	J	K	M	N	O	P	Q	S	T	V	W	X	Z	H	Y
E	E	F	G	J	K	M	N	O	P	Q	S	T	V	W	X	Z	H	X	D	R	A	U	L	I	C	B
F	F	G	J	K	M	N	O	P	Q	S	T	V	W	X	Z	H	Y	D	R	A	U	L	I	C	B	E
G	G	J	K	M	N	O	P	Q	S	T	V	W	X	Z	H	Y	D	R	A	U	L	I	C	B	E	F
H	H	Y	D	R	A	U	L	I	C	B	E	F	G	J	K	M	N	O	P	Q	S	T	V	W	X	Z
I	I	C	B	E	F	G	J	K	M	N	O	P	Q	S	T	V	W	X	Z	H	Y	D	R	A	U	L
J	J	K	M	N	O	P	Q	S	T	V	W	X	Z	H	Y	D	R	A	U	L	I	C	B	E	F	G
K	K	M	N	O	P	Q	S	T	V	W	X	Z	H	Y	D	R	A	U	L	I	C	B	E	F	G	J
L	L	I	C	B	E	F	G	J	K	M	N	O	P	Q	S	T	V	W	X	Z	H	Y	D	R	A	U
M	M	N	O	P	Q	S	T	V	W	X	Z	H	Y	D	R	A	U	L	I	C	B	E	F	G	J	K
N	N	O	P	Q	S	T	V	W	X	Z	H	Y	D	R	A	U	L	I	C	B	E	F	G	J	K	M
O	O	P	Q	S	T	V	W	X	Z	H	Y	D	R	A	U	L	I	C	B	E	F	G	J	K	M	N
P	P	Q	S	T	V	W	X	Z	H	Y	D	R	A	U	L	I	C	B	E	F	G	J	K	M	N	O
Q	Q	S	T	V	W	X	Z	H	Y	D	R	A	U	L	I	C	B	E	F	G	J	K	M	N	O	P
R	R	A	U	L	I	C	B	E	F	G	J	K	M	N	O	P	Q	S	T	V	W	X	Z	H	Y	D
S	S	T	V	W	X	Z	H	Y	D	R	A	U	L	I	C	B	E	F	G	J	K	M	N	O	P	Q
T	T	V	W	X	Z	H	Y	D	R	A	U	L	I	C	B	E	F	G	J	K	M	N	O	P	Q	S
U	U	L	I	C	B	E	F	G	J	K	M	N	O	P	Q	S	T	V	W	X	Z	H	Y	D	R	A
V	V	W	X	Z	H	Y	D	R	A	U	L	I	C	B	E	F	G	J	K	M	N	O	P	Q	S	T
W	W	X	Z	H	Y	D	R	A	U	L	I	C	B	E	F	G	J	K	M	N	O	P	Q	S	T	V
X	X	Z	H	Y	D	R	A	U	L	I	C	B	E	F	G	J	K	M	N	O	P	Q	S	T	V	W
Y	Y	D	R	A	U	L	I	C	B	E	F	G	J	K	M	N	O	P	Q	S	T	V	W	X	Z	H
Z	Z	H	Y	D	R	A	U	L	I	C	B	E	F	G	J	K	M	N	O	P	Q	S	T	V	W	X

[Plain-text letters are within the square proper]

FIGURE 12.

c. If, for mere purposes of demonstration, instead of letters within the cells of the square there are placed tallies corresponding in number with the normal frequencies of the letters occupying the respective cells, the cipher square becomes as follows (showing only the 1st three rows of the square):

ALPHABET NO.

	1	2	3	4	5	6	7	8	9	10	11	12	13	14	15	16	17	18	19	20	21	22	23	24	25	26
A	卌 ||	|||	||||	卌 |||	|||	|	卌 卌 ||	|||	||			|||	卌 |||	卌 ||	|||		卌 |	卌 ||||	|	|			|||	||	||||	卌 |||
B	|	卌 卌 ||	|||	||			|||	卌 |||	卌 ||	|||		卌 |	卌 ||||	|	|			|||	||	||||	卌 |||	卌 ||	|||	卌 |||	|||	|
C	|||	|	卌 卌 ||	|||	||			|||	卌 ||||	卌 ||	|||		卌 |	卌 ||||	|	|			|||	||	||||	卌 |||	卌 ||	|||	||||	卌 |||

FIGURE 13*a*.

d. It is obvious that here is a case wherein if two distributions pertaining to the square are isolated from the square, the χ test (matching distributions) can be applied to ascertain how the distributions should be shifted relative to each other so that they can be superimposed and made to yield a monoalphabetic composite. There is obviously one correct superimposition out of 25 possibilities. In this case, the **B** row of tallies must be displaced 5 intervals to the right in order to match it and amalgamate it with the **A** row of tallies. Thus:

A	卌|| 1	||| 2	||| 3	卌||| 4	||| 5	| 6	卌卌|| 7	||| 8	|| 9	10	11	||| 12	卌|||| 13	卌|| 14	||| 15	16	卌| 17	卌|||| 18	| 19	| 20	21	22	||| 23	|| 24	|||| 25	卌||| 26
B	卌|| 22	||| 23	卌||| 24	||| 25	| 26	| 1	卌卌|| 2	||| 3	|| 4	5	6	||| 7	卌|||| 8	卌|| [illegible]	||| 10	11	卌| 12	卌|||| 13	| 14	| 15	16	17	||| 18	|| 19	|||| 20	卌||| 21

FIGURE 13*b*.

e. Note that the amount of displacement, that is, the number of intervals the **B** sequence must be shifted to make it match the **A** sequence in Figure 13*b*, corresponds exactly to the distance between the letters A and B in the primary cipher component, which is 5 intervals. Thus:

```
 0 1 2 3 4 5
... A U L I C B ...
```

The fact that the primary plain component is in this case identical with the primary cipher component has nothing to do with the matter. *The displacement interval is being measured on the cipher component.* It is important that the student see this point very clearly. He can, if he like, prove the point by experimenting with two different primary components.

f. Assuming that a message in such a system is to be solved, the text is transcribed in rows of 26 letters. A uniliteral frequency distribution is made for each column of the transcribed text, the 26 separate distributions being compiled within a single square such as that shown in Figure 14. Such a square may be termed a *frequency distribution square.*

g. Now the vertical columns of tallies within such a distribution square constitute frequency distributions of the usual type: They show the distribution of the various cipher letters in each cipher alphabet. If there were many lines of text, all arranged in periods of 26 letters, then each column of the frequency square could be solved in the usual manner, by the application of the simple principles of monoalphabetic frequency. But what do the horizontal rows of tallies within the square represent? Is it not clear that the first such row, the **A** row, merely shows the distribution of $\mathbf{A}_c$ throughout the successive cipher alphabets? *And does not this graphic picture of the distribution of* $\mathbf{A}_c$ *correspond to the sequence of letters composing the primary plain component?* Furthermore, is it not clear that what has been said of the **A** row of tallies applies equally to the B, C, D, ... Z rows? Finally, is it not clear that the graphic pictures of all the distributions correspond to the *same* sequence of letters, except that the sequence begins with a different letter in each row? In other words, all the horizontal rows of tallies within the distribution square apply to the *same* sequence of plain-text letters, the sequences in one row merely beginning with a different letter from that with which another row begins. The sequences of letters to which the tallies apply in the various rows are merely displaced relative to one another. Now if there are sufficient data for statistical purposes in the various horizontal sequences of tallies within the distribution square, these sequences, being approximately similar, can be studied by means of the χ test *to find their relative displacements.* And in finding the latter a method is provided whereby the primary cipher component may be reconstructed, since the correct assembling of the displacement data will yield the sequence of letters constituting the primary cipher component. If the plain component is identical with the cipher component, the solution is immediately at hand; if they are different, the solution is but one step removed. Thus, there has been elaborated a method of solving this type of cipher system *without making any assumptions of values for cipher letters.*

50. Solution of a progressive-alphabet system by means of the χ test.—*a.* The following cryptogram has been enciphered according to the method indicated, by progressive, simple, uninterrupted shifting of a primary cipher component against an identical primary plain component.

CRYPTOGRAM

```
W G J J M   M M J X E   D G C O C   F T R P B   M I I I K   Z R Y N N
B U F R W   W W W Y O   I H F J K   O K H T T   A Z C L J   E P P F R
W C K O O   F F F G E   P Q R Y Y   I W X M X   U D I P F   E X M L L
W F K G Y   P B B X C   H B F Y I   E T X H F   B I V D I   P N X I V
R P W T M   G I M P T   E C J B O   K V B U Q   G V G F F   F K L Y Y
C K B I W   X M X U D   I P F F U   Y N V S S   I H R M H   Y Z H A U
Q W G K T   I U X Y J   J A O W Z   O C F T R   P P O Q U   S G Y C X
V C X U C   J L M L L   Y E K F F   Z V Q J Q   S I Y S P   D S B B J
U A H Y N   W L O C X   S D Q V C   Y V S I L   I W N J O   O M A Q S
L W Y J G   T V P Q K   P K T L H   S R O O N   I C F E V   M N V W N
B N E H A   M R C R O   V S T X E   N H P V B   T W K U Q   I O C A V
W B R Q N   F J V N R   V D O P U   Q R L K Q   N F F F Z   P H U R V
W L X G S   H Q W H P   J B C N N   J Q S O Q   O R C B M   R R A O N
R K W U H   Y Y C I W   D G S J C   T G P G R   M I Q M P   S G C T N
M F G J X   E D G C O   P T G P W   Q Q V Q I   W X T T T   C O J V A
A A B W M   X I H O W   H D E Q U   A I N F K   F W H P J   A H Z I T
W Z K F E   X S R U Y   Q I O V R   E R D J V   D K H I R   Q W E D G
E B Y B M   L A B J V   T G F F G   X Y I V G   R J Y E K   F B E P B
J O U A H   C U G Z L   X I A J K   W D V T Y   B F R U C   C C U Z Z
I N N D F   R J F M B   H Q L X H   M H Q Y Y   Y M W Q V   C L I P T
W T J Y Q   B Y R L I   T U O U S   R C D C V   W D G I G   G U B H J
V V P W A   B U J K N   F P F Y W   V Q Z Q F   L H T W J   P D R X Z
O W U S S   G A M H N   C W H S W   W L R Y Q   Q U S Z V   D N X A N
V N K H F   U C V V S   S S P L Q   U P C V V   V W D G S   J O G T C
H D E V Q   S I J P H   Q J A W F   R I Z D W   X X H C X   Y C T M G
U S E S N   D S B B K   R L V W R   V Z E E P   P P A T O   I A N E E
E E J N R   C Z B T B   L X P J J   K A P P M   J E G I K   R T G F F
H P V V V   Y K J E F   H Q S X J   Q D Y V Z   G R R H Z   Q L Y X K
X A Z O W   R R X Y K   Y G M G Z   B Y N V H   Q B R V F   E F Q L L
W Z E Y L   J E R O Q   S O Q K O   M W I O G   M B K F F   L X D X T
L W I L P   Q S E D Y   I O E M O   I B J M L   N N S Y K   X J Z J M
L C Z B M   S D J W Q   X T J V L   F I R N R   X H Y B D   B J U F I
R J I C T   U U U S K   K W D V M   F W T T J   K C K C G   C V S A G
Q B C J M   E B Y N V   S S J K S   D C B D Y   F P P V F   D W Z M T
B P V T T   C G B V T   Z K H Q D   D R M E Z   O O
```

b. The message is transcribed in lines of 26 letters, since that is the total number of secondary alphabets in the system. The transcribed text is shown below:

	1	2	3	4	5	6	7	8	9	10	11	12	13	14	15	16	17	18	19	20	21	22	23	24	25	26
1	W	G	J	J	M	M	M	J	X	E	D	G	C	O	C	F	T	R	P	B	M	I	I	I	K	Z
2	R	Y	N	N	B	U	F	R	W	W	W	W	Y	O	I	H	F	J	K	O	K	H	T	T	A	Z
3	C	L	J	E	P	P	F	R	W	C	K	O	O	F	F	F	G	E	P	Q	R	Y	Y	I	W	X
4	M	X	U	D	I	P	F	E	X	M	L	L	W	F	K	G	Y	P	B	B	X	C	H	B	F	Y
5	I	E	T	X	H	F	B	I	V	D	I	P	N	X	I	V	R	P	W	T	M	G	I	M	P	T
6	E	C	J	B	O	K	V	B	U	Q	G	V	G	F	F	F	K	L	Y	Y	C	K	B	I	W	X
7	M	X	U	D	I	P	F	F	U	Y	N	V	S	S	I	H	R	M	H	Y	Z	H	A	U	Q	W
8	G	K	T	I	U	X	Y	J	J	A	O	W	Z	O	C	F	T	R	P	P	O	Q	U	S	G	Y
9	C	X	V	C	X	U	C	J	L	M	L	L	Y	E	K	F	F	Z	V	Q	J	Q	S	I	Y	S
10	P	D	S	B	B	J	U	A	H	Y	N	W	L	O	C	X	S	D	Q	V	C	Y	V	S	I	L
11	I	W	N	J	O	O	M	A	Q	S	L	W	Y	J	G	T	V	P	Q	K	P	K	T	L	H	S
12	R	O	O	N	I	C	F	E	V	M	N	V	W	N	B	N	E	H	A	M	R	C	R	O	V	S
13	T	X	E	N	H	P	V	B	T	W	K	U	Q	I	O	C	A	V	W	B	R	Q	N	F	J	V
14	N	R	V	D	O	P	U	Q	R	L	K	Q	N	F	F	F	Z	P	H	U	R	V	W	L	X	G
15	S	H	Q	W	H	P	J	B	C	N	N	J	Q	S	O	Q	O	R	C	B	M	R	R	A	O	N
16	R	K	W	U	H	Y	Y	C	I	W	D	G	S	J	C	T	G	P	G	R	M	I	Q	M	P	S
17	G	C	T	N	M	F	G	J	X	E	D	G	C	O	P	T	G	P	W	Q	Q	V	Q	I	W	X
18	T	T	T	C	O	J	V	A	A	A	B	W	M	X	I	H	O	W	H	D	E	Q	U	A	I	N
19	F	K	F	W	H	P	J	A	H	Z	I	T	W	Z	K	F	E	X	S	R	U	Y	Q	I	O	V
20	R	E	R	D	J	V	D	K	H	I	R	Q	W	E	D	G	E	B	Y	B	M	L	A	B	J	V
21	T	G	F	F	G	X	Y	I	V	G	R	J	Y	E	K	F	B	E	P	B	J	O	U	A	H	C
22	U	G	Z	L	X	I	A	J	K	W	D	V	T	Y	B	F	R	U	C	C	C	U	Z	Z	I	N
23	N	D	F	R	J	F	M	B	H	Q	L	X	H	M	H	Q	Y	Y	Y	M	W	Q	V	C	L	I
24	P	T	W	T	J	Y	Q	B	Y	R	L	I	T	U	O	U	S	R	C	D	C	V	W	D	G	I
25	G	G	U	B	H	J	V	V	P	W	A	B	U	J	K	N	F	P	F	Y	W	V	Q	Z	Q	F
26	L	H	T	W	J	P	D	R	X	Z	O	W	U	S	S	G	A	M	H	N	C	W	H	S	W	W
27	L	R	Y	Q	Q	U	S	Z	V	D	N	X	A	N	V	N	K	H	F	U	C	V	V	S	S	S
28	P	L	Q	U	P	C	V	V	V	W	D	G	S	J	O	G	T	C	H	D	E	V	Q	S	I	J
29	P	H	Q	J	A	W	F	R	I	Z	D	W	X	X	H	C	X	Y	C	T	M	G	U	S	E	S
30	N	D	S	B	B	K	R	L	V	W	R	V	Z	E	E	P	P	P	A	T	O	I	A	N	E	E
31	E	E	J	N	R	C	Z	B	T	B	L	X	P	J	J	K	A	P	P	M	J	E	G	I	K	R
32	T	G	F	F	H	P	V	V	V	Y	K	J	E	F	H	Q	S	X	J	Q	D	Y	V	Z	G	R
33	R	H	Z	Q	L	Y	X	K	X	A	Z	O	W	R	R	X	Y	K	Y	G	M	G	Z	B	Y	N
34	V	H	Q	B	R	V	F	E	F	Q	L	L	W	Z	E	Y	L	J	E	R	O	Q	S	O	Q	K
35	O	M	W	I	O	G	M	B	K	F	F	L	X	D	X	T	L	W	I	L	P	Q	S	E	D	Y
36	I	O	E	M	O	I	B	J	M	L	N	N	S	Y	K	X	J	Z	J	M	L	C	Z	B	M	S
37	D	J	W	Q	X	T	J	V	L	F	I	R	N	R	X	H	Y	B	D	B	J	U	F	I	R	J
38	I	C	T	U	U	U	S	K	K	W	D	V	M	F	W	T	T	J	K	C	K	C	G	C	V	S
39	A	G	Q	B	C	J	M	E	B	Y	N	V	S	S	J	K	S	D	C	B	D	Y	F	P	P	V
40	F	D	W	Z	M	T	B	P	V	T	T	C	G	B	V	T	Z	K	H	Q	D	D	R	M	E	Z
41	O	O																								

c. A frequency distribution square is then compiled, each column of the text forming a separate distribution in columnar form in the square. The latter is shown in figure 14.

	1	2	3	4	5	6	7	8	9	10	11	12	13	14	15	16	17	18	19	20	21	22	23	24	25	26	N
A																											25
B																											43
C																											45
D																											34
E																											35
F																											51
G																											39
H																											38
I																											45
J																											50
K																											37
L																											33
M																											37
N																											34
O																											38
P																											43
Q																											45
R																											46
S																											39
T																											39
U																											33
V																											53
W																											52
X																											37
Y																											44
Z																											27
	1	2	3	4	5	6	7	8	9	10	11	12	13	14	15	16	17	18	19	20	21	22	23	24	25	26	1,042

FIGURE 14.

d. The χ test will now be applied to the horizontal rows of tallies in the distribution square, in accordance with the theory set forth in paragraph 49*g*. Since this test is purely statistical in character and becomes increasingly reliable as the size of the distributions increases, it is best to start by working with the two distributions having the greatest total numbers of tallies. These are the V and W distributions, with 53 and 52 occurrences, respectively. The results of three relative displacements of these two distributions are shown below, labeled "First test," "Second test," and "Third test."

FIRST TEST

f_V	1	0	2	0	0	2	6	4	8	0	0	7	0	0	2	1	1	1	1	1	0	6	4	0	2	4	$N_V=53$
	1	2	3	4	5	6	7	8	9	10	11	12	13	14	15	16	17	18	19	20	21	22	23	24	25	26	
f_W	24	25	26	1	2	3	4	5	6	7	8	9	10	11	12	13	14	15	16	17	18	19	20	21	22	23	
	0	4	2	1	1	5	3	0	1	0	0	2	8	1	7	6	0	1	0	0	2	3	0	2	1	2	$N_W=52$
f_Vf_W	0	0	4	0	0	10	18	0	8	0	0	14	0	0	14	6	0	1	0	0	0	18	0	0	2	8	$\Sigma f_Vf_W=103$

$$\frac{\Sigma f_Vf_W}{N_VN_W}=\frac{103}{2756}=.037$$

SECOND TEST

f_V	1	0	2	0	0	2	6	4	8	0	0	7	0	0	2	1	1	1	1	1	0	6	4	0	2	4	$N_V=53$
	1	2	3	4	5	6	7	8	9	10	11	12	13	14	15	16	17	18	19	20	21	22	23	24	25	26	
f_W	18	19	20	21	22	23	24	25	26	1	2	3	4	5	6	6	8	9	10	11	12	13	14	15	16	17	
	2	3	0	2	1	2	0	4	2	1	1	5	3	0	1	0	0	2	8	1	7	6	0	1	0	0	$N_W=52$
f_Vf_W	2	0	0	0	0	4	0	16	16	0	0	35	0	0	2	0	0	2	8	1	0	36	0	0	0	0	$\Sigma f_Vf_W=122$

$$\frac{\Sigma f_Vf_W}{N_VN_W}=\frac{122}{2756}=.044$$

THIRD TEST

f_V	1	0	2	0	0	2	6	4	8	0	0	7	0	0	2	1	1	1	1	1	0	6	4	0	2	4	$N_V=53$
	1	2	3	4	5	6	7	8	9	10	11	12	13	14	15	16	17	18	19	20	21	22	23	24	25	26	
f_W	4	5	6	7	8	9	10	11	12	13	14	15	16	17	18	19	20	21	22	23	24	25	26	1	2	3	
	3	0	1	0	0	2	8	1	7	6	0	1	0	0	2	3	0	2	1	2	0	4	2	1	1	5	$N_W=52$
f_Vf_W	3	0	2	0	0	4	48	4	56	0	0	7	0	0	4	3	0	2	1	2	0	24	8	0	2	20	$\Sigma f_Vf_W=190$

$$\frac{\Sigma f_Vf_W}{N_VN_W}=\frac{190}{2756}=.069$$

e. Since the last of the three foregoing tests gives a value somewhat better than the expected .0667, it looks as though the correct position of the W distribution with reference to the V distribution has been found. In practice, several more tests would be made to insure that other close approximations to .0667 will not be found, but these will here be omitted. The test indicates that the primary cipher component has the letters V and W in these positions: $\overset{1}{\text{V}}\ \overset{2}{.}\ \overset{3}{.}\ \overset{4}{\text{W}}$, since the correct superimposition requires that the 4th cell of the W distribution must be placed under the 1st cell of the V distribution (see the last superimposition above).

f. The next best distribution with which to proceed is the F distribution, with 51 occurrences. Paralleling the procedure outlined in paragraph 43, and for the same reasons, the F sequence is matched against the W and V sequences separately and then against both W and V sequences at their correct superimposition. The following shows the correct relative positions of the three distributions:

f_V	1	0	2	0	0	2	6	4	8	0	0	7	0	0	2	1	1	1	1	1	0	6	4	0	2	4	$N_V=53$
	1	2	3	4	5	6	7	8	9	10	11	12	13	14	15	16	17	18	19	20	21	22	23	24	25	26	
f_F	8	9	10	11	12	13	14	15	16	17	18	19	20	21	22	23	24	25	26	1	2	3	4	5	6	7	
	1	1	2	1	0	0	6	3	9	3	0	2	0	0	0	2	1	1	1	2	0	4	2	0	3	7	$N_F=51$
f_Vf_F	1	0	4	0	0	0	36	12	72	0	0	14	0	0	0	2	1	1	1	2	0	24	8	0	6	28	$\Sigma f_Vf_F=212$

$$\frac{\Sigma f_V f_F}{N_V N_F}=\frac{212}{2,703}=.078$$

---

f_W	1	1	5	3	0	1	0	0	2	8	1	7	6	0	1	0	0	2	3	0	2	1	2	0	4	2	$N_W=52$
	1	2	3	4	5	6	7	8	9	10	11	12	13	14	15	16	17	18	19	20	21	22	23	24	25	26	
f_F	5	6	7	8	9	10	11	12	13	14	15	16	17	18	19	20	21	22	23	24	25	26	1	2	3	4	
	0	3	7	1	1	2	1	0	0	6	3	9	3	0	2	0	0	0	2	1	1	1	2	0	4	2	$N_F=51$
f_Wf_F	0	3	35	[illegible]	0	2	0	0	0	48	3	63	18	0	2	0	0	0	6	0	2	1	4	0	16	4	$\Sigma f_Wf_F=210$

$$\frac{\Sigma f_W f_F}{N_W N_F}=\frac{210}{2,703}=.078$$

$f(V+W)$	4	0	3	0	0	4	14	5	15	6	0	8	0	0	4	4	1	3	2	3	0	10	6	1	3	9	$N_{V+W}=105$
	1	2	3	4	5	6	7	8	9	10	11	12	13	14	15	16	17	18	19	20	21	22	23	24	25	26	
f_F	8	9	10	11	12	13	14	15	16	17	18	19	20	21	22	23	24	25	26	1	2	3	4	5	6	7	
	1	1	2	1	0	0	6	3	9	3	0	2	0	0	0	2	1	1	1	2	0	4	2	0	3	7	$N_F=51$
$f(V+W)f_F$	4	0	6	0	0	0	84	15	35	18	0	16	0	0	0	8	1	3	2	6	0	40	12	0	9	63	$\Sigma f_{(V+W)}f_F=422$

$$\frac{\Sigma f_{(V+W)} f_F}{N_{(V+W)} N_F}=\frac{422}{5,355}=.079$$

The test yields the sequence

1	2	3	4	5	6	7	8
V	.	.	W	.	.	.	F

g. The process is continued in the foregoing manner until the entire primary cipher component has been reconstructed. It is obvious that as the work progresses the cryptanalyst is forced to employ smaller and smaller distributions, so that statistically the results are apt to become less and less certain. But to counterbalance this there is the fact that the number of possible superimpositions becomes progressively smaller as the work progresses. For example, at the commencement of operations the number of possible points for superimposing a second sequence against the first is 25; after the relative positions of 5 distributions have been ascertained and a 6th distribution is to be placed in the primary sequence being reconstructed, there are 21 possible positions; after the relative positions of 20 distributions have been ascertained, there are only 6 possible positions for the 21st distribution, and so on.

h. In the foregoing case the completely reconstructed primary cipher component is as follows:

1	2	3	4	5	6	7	8	9	10	11	12	13	14	15	16	17	18	19	20	21	22	23	24	25	26
V	A	L	W	N	O	X	F	B	P	Y	R	C	Q	Z	I	G	S	E	H	T	D	J	U	M	K

Since it was stated that the problem involves identical primary components, both components are now at hand.

i. Of course, it is probable that in practical work the process of matching distributions would be interrupted soon after the positions of only a few letters in the primary component had been ascertained. For by trying partially reconstructed sequences on the cipher text the skeletons of some words would begin to show. By filling in these skeletons with the words suggested by them, the process of reconstructing the components is much facilitated and hastened.

j. The components having been reconstructed, only a moment or two is necessary to ascertain their initial position in enciphering the message. It is only necessary to juxtapose the two components so as to give "good" values for any one of the vertical distributions of Figure 14. This then gives the juxtaposition of the components for that column, and the rest follows very easily for the plain text may now be obtained by direct use of the components. The plain text of the message is as follows:

	1	2	3	4	5	6	7	8	9	10	11	12	13	14	15	16	17	18	19	20	21	22	23	24	25	26
1	W	G	J	J	M	M	M	J	X	E	D	G	C	O	C	F	T	R	P	B	M	I	I	I	K	Z
	W	I	T	H	T	H	E	I	M	P	R	O	V	E	M	E	N	T	S	I	N	T	H	E	A	I
2	R	Y	N	N	B	U	F	R	W	W	W	W	Y	O	I	H	F	J	K	O	K	H	T	T	A	Z
	R	P	L	A	N	E	A	N	D	T	H	E	M	E	A	N	S	O	F	C	O	M	M	U	N	I
3	C	L	J	E	P	P	F	R	W	C	K	O	O	F	F	F	G	E	P	Q	R	Y	Y	I	W	X
	C	A	T	I	O	N	A	N	D	W	I	T	H	T	H	E	V	A	S	T	S	I	Z	E	O	F
4	M	X	U	D	I	P	F	E	X	M	L	L	W	F	K	G	Y	P	B	B	X	C	H	B	F	Y
	M	O	D	E	R	N	A	R	M	I	E	S	S	T	R	A	T	E	G	I	C	S	U	R	P	R
5	I	E	T	X	H	F	B	I	V	D	I	P	N	X	I	V	R	P	W	T	M	G	I	M	P	T
	I	S	E	W	I	L	L	B	E	C	O	M	E	H	A	R	D	E	R	A	N	D	H	A	R	D
6	E	C	J	B	O	K	V	B	U	Q	G	V	G	F	F	F	K	L	Y	Y	C	K	B	I	W	X
	E	R	T	O	A	T	T	A	I	N	X	I	N	T	H	E	P	R	E	S	E	N	C	E	O	F
7	M	X	U	D	I	P	F	F	U	Y	N	V	S	S	I	H	R	M	H	Y	Z	H	A	U	Q	W
	M	O	D	E	R	N	A	V	I	A	T	I	O	N	A	N	D	F	A	S	T	M	O	V	I	N
8	G	K	T	I	U	X	Y	J	J	A	O	W	Z	O	C	F	T	R	P	P	O	Q	U	S	G	Y
	G	M	E	C	H	A	N	I	Z	E	D	E	L	E	M	E	N	T	S	G	R	E	A	T	E	R
9	C	X	V	C	X	U	C	J	L	M	L	L	Y	E	K	F	F	Z	V	Q	J	Q	S	I	Y	S
	C	O	M	P	L	E	X	I	T	I	E	S	M	O	R	E	S	U	B	T	L	E	D	E	C	E
10	P	D	S	B	B	J	U	A	H	Y	N	W	L	O	C	X	S	D	Q	V	C	Y	V	S	I	L
	P	T	I	O	N	S	S	T	R	A	T	E	G	E	M	S	A	N	D	F	E	I	N	T	S	W
11	I	W	N	J	O	O	M	A	Q	S	L	W	Y	J	G	T	V	P	Q	K	P	K	T	L	H	S
	I	L	L	H	A	V	E	T	O	B	E	E	M	P	L	O	Y	E	D	X	I	N	M	O	D	E
12	R	O	O	N	I	C	F	E	V	M	N	V	W	N	B	N	E	H	A	M	R	C	R	O	V	S
	R	N	W	A	R	F	A	R	E	I	T	I	S	S	T	I	L	L	P	O	S	S	I	B	L	E
13	T	X	E	N	H	P	V	B	T	W	K	U	Q	I	O	C	A	V	W	B	R	Q	N	F	J	V
	T	O	G	A	I	N	T	A	C	T	I	C	A	L	S	U	R	P	R	I	S	E	B	Y	M	A
14	N	R	V	D	O	P	U	Q	R	L	K	Q	N	F	F	F	Z	P	H	U	R	V	W	L	X	G
	N	Y	M	E	A	N	S	X	W	H	I	L	E	T	H	E	M	E	A	N	S	O	F	O	B	S
15	S	H	Q	W	H	P	J	B	C	N	N	J	Q	S	O	Q	O	R	C	B	M	R	R	A	O	N
	S	E	R	V	I	N	G	A	N	D	T	R	A	N	S	M	I	T	T	I	N	G	I	N	F	O

	1	2	3	4	5	6	7	8	9	10	11	12	13	14	15	16	17	18	19	20	21	22	23	24	25	26
16	R	K	W	U	H	Y	Y	C	I	W	D	G	S	J	C	T	G	P	G	R	M	I	Q	M	P	S
	R	M	A	T	I	O	N	O	F	T	R	O	O	P	M	O	V	E	M	E	N	T	S	A	R	E
17	G	C	T	N	M	F	G	J	X	E	D	G	C	O	P	T	G	P	W	Q	Q	V	Q	I	W	X
	G	R	E	A	T	L	Y	I	M	P	R	O	V	E	D	O	V	E	R	T	H	O	S	E	O	F
18	T	T	T	C	O	J	V	A	A	A	B	W	M	X	I	H	O	W	H	D	E	Q	U	A	I	N
	T	H	E	P	A	S	T	T	H	E	M	E	C	H	A	N	I	C	A	L	M	E	A	N	S	O
19	F	K	F	W	H	P	J	A	H	Z	I	T	W	Z	K	F	E	X	S	R	U	Y	Q	I	O	V
	F	M	O	V	I	N	G	T	R	O	O	P	S	A	R	E	L	I	K	E	W	I	S	E	F	A
20	R	E	R	D	J	V	D	K	H	I	R	Q	W	E	D	G	E	B	Y	B	M	L	A	B	J	V
	R	S	P	E	E	D	I	E	R	X	A	L	S	O	F	A	L	S	E	I	N	F	O	R	M	A
21	T	G	F	F	G	X	Y	I	V	G	R	J	Y	E	K	F	B	E	P	B	J	O	U	A	H	C
	T	I	O	N	C	A	N	B	E	F	A	R	M	O	R	E	E	A	S	I	L	Y	A	N	D	Q
22	U	G	Z	L	X	I	A	J	K	W	D	V	T	Y	B	F	R	U	C	C	C	U	Z	Z	I	N
	U	I	C	K	L	Y	D	I	S	T	R	I	B	U	T	E	D	X	T	H	E	L	E	S	S	O
23	N	D	F	R	J	F	M	B	H	Q	L	X	H	M	H	Q	Y	Y	Y	M	W	Q	V	C	L	I
	N	T	O	B	E	L	E	A	R	N	E	D	F	R	O	M	T	H	E	O	P	E	N	I	N	G
24	P	T	W	T	J	Y	Q	B	Y	R	L	I	T	U	O	U	S	R	C	D	C	V	W	D	G	I
	P	H	A	S	E	O	F	A	L	L	E	N	B	Y	S	B	A	T	T	L	E	O	F	M	E	G
25	G	G	U	B	H	J	V	V	P	W	A	B	U	J	K	N	F	P	F	Y	W	V	Q	Z	Q	F
	G	I	D	O	I	S	T	H	A	T	S	U	R	P	R	I	S	E	I	S	P	O	S	S	I	B
26	L	H	T	W	J	P	D	R	X	Z	O	W	U	S	S	G	A	M	H	N	C	W	H	S	W	W
	L	E	E	V	E	N	I	N	M	O	D	E	R	N	W	A	R	F	A	R	E	B	U	T	O	N
27	L	R	Y	Q	Q	U	S	Z	V	D	N	X	A	N	V	N	K	H	F	U	C	V	V	S	S	S
	L	Y	B	Y	P	E	R	F	E	C	T	D	I	S	C	I	P	L	I	N	E	O	N	T	H	E
28	P	L	Q	U	P	C	V	V	V	W	D	G	S	J	O	G	T	C	H	D	E	V	Q	S	I	J
	P	A	R	T	O	F	T	H	E	T	R	O	O	P	S	A	N	D	A	L	M	O	S	T	S	U
29	P	H	Q	J	A	W	F	R	I	Z	D	W	X	X	H	C	X	Y	C	T	M	G	U	S	E	S
	P	E	R	H	U	M	A	N	F	O	R	E	T	H	O	U	G	H	T	A	N	D	A	T	T	E
30	N	D	S	B	B	K	R	L	V	W	R	V	Z	E	E	P	P	P	A	T	O	I	A	N	E	E
	N	T	I	O	N	T	O	D	E	T	A	I	L	O	N	T	H	E	P	A	R	T	O	F	T	H
31	E	E	J	N	R	C	Z	B	T	B	L	X	P	J	J	K	A	P	P	M	J	E	G	I	K	R
	E	S	T	A	F	F	B	A	C	K	E	D	U	P	B	Y	R	E	S	O	L	U	T	E	A	C
23	T	G	F	F	H	P	V	V	V	Y	K	J	E	F	H	Q	S	X	J	Q	D	Y	V	Z	G	R
	T	I	O	N	I	N	T	H	E	A	I	R	X	T	O	M	A	I	N	T	A	I	N	S	E	C
33	R	H	Z	Q	L	Y	X	K	X	A	Z	O	W	R	R	X	Y	K	Y	G	M	G	Z	B	Y	N
	R	E	C	Y	M	O	V	E	M	E	N	T	S	M	U	S	T	B	E	U	N	D	E	R	C	O
34	V	H	Q	B	R	V	F	E	F	Q	L	L	W	Z	E	Y	L	J	E	R	O	Q	S	O	Q	K
	V	E	R	O	F	D	A	R	K	N	E	S	S	A	N	D	C	O	V	E	R	E	D	B	I	V
35	O	M	W	I	O	G	M	B	K	F	F	L	X	D	X	T	L	W	I	L	P	Q	S	E	D	Y
	O	U	A	C	A	R	E	A	S	M	U	S	T	B	E	O	C	C	U	P	I	E	D	D	U	R

	1	2	3	4	5	6	7	8	9	10	11	12	13	14	15	16	17	18	19	20	21	22	23	24	25	26
36	I	O	E	M	O	I	B	J	M	L	N	N	S	Y	K	X	J	Z	J	M	L	C	Z	B	M	S
	I	N	G	D	A	Y	L	I	G	H	T	H	O	U	R	S	X	U	N	O	B	S	E	R	V	E
37	D	J	W	Q	X	T	J	V	L	F	I	R	N	R	X	H	Y	B	D	B	J	U	F	I	R	J
	D	D	A	Y	L	I	G	H	T	M	O	V	E	M	E	N	T	S	W	I	L	L	R	E	Q	U
38	I	C	T	U	U	U	S	K	K	W	D	V	M	F	W	T	T	J	K	C	K	C	G	C	V	S
	I	R	E	T	H	E	R	E	S	T	R	I	C	T	I	O	N	O	F	H	O	S	T	I	L	E
39	A	G	Q	B	C	J	M	E	B	Y	N	V	S	S	J	K	S	D	C	B	D	Y	F	P	P	V
	A	I	R	O	B	S	E	R	V	A	T	I	O	N	B	Y	A	N	T	I	A	I	R	C	R	A
40	F	D	W	Z	M	T	B	P	V	T	T	C	G	B	V	T	Z	K	H	Q	D	D	R	M	E	Z
	F	T	A	R	T	I	L	L	E	R	Y	A	N	D	C	O	M	B	A	T	A	V	I	A	T	I
41	O	O																								
	O	N																								

WITH THE IMPROVEMENTS IN THE AIRPLANE AND THE MEANS OF COMMUNICATION AND WITH THE VAST SIZE OF MODERN ARMIES STRATEGIC SURPRISE WILL BECOME HARDER AND HARDER TO ATTAIN X IN THE PRESENCE OF MODERN AVIATION AND FAST MOVING MECHANIZED ELEMENTS GREATER COMPLEXITIES MORE SUBTLE DECEPTIONS STRATEGEMS AND FEINTS WILL HAVE TO BE EMPLOYED X IN MODERN WARFARE IT IS STILL POSSIBLE TO GAIN TACTICAL SURPRISE BY MANY MEANS X WHILE THE MEANS OF OBSERVING AND TRANSMITTING INFORMATION OF TROOP MOVEMENTS ARE GREATLY IMPROVED OVER THOSE OF THE PAST THE MECHANICAL MEANS OF MOVING TROOPS ARE LIKEWISE FAR SPEEDIER X ALSO FALSE INFORMATION CAN BE FAR MORE EASILY AND QUICKLY DISTRIBUTED X THE LESSON TO BE LEARNED FROM THE OPENING PHASE OF ALLENBYS BATTLE OF MEGGIDO IS THAT SURPRISE IS POSSIBLE EVEN IN MODERN WARFARE BUT ONLY BY PERFECT DISCIPLINE ON THE PART OF THE TROOPS AND ALMOST SUPERHUMAN FORETHOUGHT AND ATTENTION TO DETAIL ON THE PART OF THE STAFF BACKED UP BY RESOLUTE ACTION IN THE AIR X TO MAINTAIN SECRECY MOVE- MENTS MUST BE UNDER COVER OF DARKNESS AND COVERED BIVOUAC AREAS MUST BE OCCUPIED DURING DAYLIGHT HOURS X UNOBSERVED DAYLIGHT MOVEMENTS WILL REQUIRE THE RESTRIC- TION OF HOSTILE AIR OBSERVATION BY ANTIAIRCRAFT ARTILLERY AND COMBAT AVIATION.

k. The student should clearly understand the real nature of the matching process employed to such good advantage in this problem. In practically all the previous cases frequency distributions were made of *cipher letters* occurring in a cryptogram, and the tallies in those distributions represented the actual occurrences of cipher letters. Furthermore, when these distributions were compared or matched, what were being compared were actually cipher alphabets. That is, the text was arranged in a certain way, so that letters belonging to the same cipher alphabet actually fell within the same column and the frequency distribution for a specific cipher alphabet was made by tabulating the letters in that column. Then if any distributions were to be compared, usually the entire distribution applicable to one cipher alphabet was compared with the entire distribution applying to another cipher alphabet. But in the problem just completed, what were compared in reality were not frequency distributions applying to the *columns* of the cipher text as transcribed on p. 83, but graphic representations of the variations in the frequencies of *plain-text letters falling in identical sequences, the identities of these plain-text letters being unknown for the moment.* Only after the reconstruction has been completed do their identities become known, when the plain text of the cryptogram is established.

51. Alternative method of solution.—*a.* The foregoing method of solution is, of course, almost entirely statistical in nature. There is, however, another method of attack which should be brought to notice because in some cases the statistical method, involving the study of relatively large distributions, may not be feasible for lack of sufficient text. Yet in these cases there may be sufficient data in the respective alphabets to permit of some assumptions of values of cipher letters, or there may be good grounds for applying the probable-word method. The present paragraph will therefore deal with a method of solving progressive cipher systems which is based upon the application of the principles of indirect symmetry to certain phenomena arising from the mechanics of the progressive encipherment method itself.

b. Take the two sequences below and encipher the phrase FIRST BATTALION by the progressive method, sliding the cipher component to the left one interval after each encipherment.

COMPONENTS

Plain	H	Y	D	R	A	U	L	I	C	B	E	F	G	J	K	M	N	O	P	Q	S	T	V	W	X	Z
Cipher	F	B	P	Y	R	C	Q	Z	I	G	S	E	H	T	D	J	U	M	K	V	A	L	W	N	O	X

MESSAGE

	1	2	3	4	5	6	7	8	9	10	11	12	13	14
Plain	F	I	R	S	T	B	A	T	T	A	L	I	O	N
Cipher	E	I	C	N	X	D	S	P	Y	T	U	K	Y	Y

c. Certain letters are repeated in both plain text and cipher text. Consider the former. There are two I's, three T's, and two A's. Their encipherments are isolated below, for convenience in study.

	F	I	R	S	T	B	A	T	T	A	L	I	O	N	
	1	2	3	4	5	6	7	8	9	10	11	12	13	14	
Plain	.	I	.	.	.	.	.	.	.	.	.	I	.	.	(1)
Cipher	.	I	.	.	.	.	.	.	.	.	.	K	.	.	(2)
Plain	.	.	.	.	T	.	.	T	T	.	.	.	.	.	(3)
Cipher	.	.	.	.	X	.	.	P	Y	.	.	.	.	.	(4)
Plain	.	.	.	.	.	.	A	.	.	A	.	.	.	.	(5)
Cipher	.	.	.	.	.	.	S	.	.	T	.	.	.	.	(6)

The two I's in line (1) are 10 letters apart; reference to the cipher component will show that the interval between the cipher equivalent of the first I_p (which happens to be I_c) and the second I_p (which is K_c) is 10. Consideration of the mechanics of the enciphering system soon shows why this is so: since the cipher component is displaced one step with each encipherment, two identical letters *n* intervals apart in the plain text must yield cipher equivalents which are *n* intervals apart in the cipher component. Examination of the data in lines (3) and (4), (5) and (6) will confirm this finding. Consequently, it would appear that in such a system the successful application of the probable-word method of attack, coupled within indirect symmetry, can quickly lead to the reconstruction of the cipher component.

d. Now consider the repeated cipher letters in the example under *b*. There happens to be only two cases of repetition, both involving Y's. Thus:

1	2	3	4	5	6	7	8	9	10	11	12	13	14
.	.	.	.	.	.	.	.	T	.	.	.	O	N
.	.	.	.	.	.	.	.	Y	.	.	.	Y	Y

Reference to the plain component will show that the plain-text letters represented by the three Y's appear in the order N O . . . T, that is, reversed with respect to their order in the plain text. But the intervals between these letters is correct. Again a consideration of the mechanics of the enciphering system shows why this is so: since the cipher component is displaced one step with each encipherment, two identical letters *n* intervals apart in the cipher text must represent plain-text letters which are *n* intervals apart in the plain component. In the present case the direction in which these letters run in the plain component is opposite to that in which the cipher component is displaced. That is, if the cipher component is displaced toward the left, the values obtained from a study of repeated plain-text letters give letters which coincide in sequence (interval and direction) with the same letters in the cipher component; the values obtained from a study of repeated cipher-text letters give letters the order of which must be reversed in order to make these letters coincide in sequence (interval and direction) with the same letters in the plain component. If the cipher component is displaced toward the right, this relationship is merely reversed: the values obtained from a study of the repeated plain-text letters must be reversed in their order when placing them in the cipher component; those yielded by a study of the repeated cipher-text letters are inserted in the plain component in their original order.

e. Of course, if the primary components are identical sequences the data from the two sources referred to in subparagraphs *c* and *d* need not be kept separate but can be combined and made to yield the primary component very quickly.

f. With the foregoing principles as background, and given the following message, which is assumed to begin with COMMANDING GENERAL FIRST ARMY (probable-word method of attack), the data yielded by this assumed text are shown in Figure 15.

MESSAGE

IKMKI LIDOL WLPNM VWPXW DUFFT
FNIIG XGAMX CADUV AZVIS YNUNL etc., etc.

	1	2	3	4	5	6	7	8	9	10	11	12	13	14	15	16	17	18	19	20	21	22	23	24	25	26
Assumed plain text	C	O	M	M	A	N	D	I	N	G	G	E	N	E	R	A	L	F	I	R	S	T	A	R	M	Y
Cipher	I	K	M	K	I	L	I	D	O	L	W	L	P	N	M	V	W	P	X	W	D	U	F	F	T	F

	A	B	C	D	E	F	G	H	I	J	K	L	M	N	O	P	Q	R	S	T	U	V	W	X	Y	Z
1			I																							
2															K											
3													M													
4													K													
5	I																									
6														L												
7				I																						
8									D																	
9														O												
10							L																			
11							W																			
12					L																					
13														P												
14					N																					
15																		M								
16	V																									
17												W														
18						P																				
19									X																	
20																		W								
21																			D							
22																				U						
23	F																									
24																		F								
25													T													
26																									F	

FIGURE 15.

Analysis of the data afforded by Figure 15, in conjunction with the principles of indirect symmetry, yields the following partial components:

	1	2	3	4	5	6	7	8	9	10	11	12	13	14	15	16	17	18	19	20	21	22	23	24	25	26
Plain	A	.	L	I	C	.	E	F	G	.	.	M	N	O	.	.	S	.	.	.	.	.	.	Y	D	R
Cipher	.	.	M	K	V	.	L	W	N	O	.	F	.	P	.	.	.	.	.	I	.	.	.	.	T	.
	D	.	.	.	.	.	.	.	.	.	.	X														

Setting the two partial components into juxtaposition so that $C_p = I_c$ (first encipherment) the 8th value, $I_p = D_c$, gives the position of D in the cipher component and permits the addition of X to it, these being two letters which until now could not be placed into position in the cipher component. With these two partial sequences it becomes possible now to decipher many other letters in the message, gaps being filled in from the context. For example, the first few letters after ARMY decipher as follows:

	1	2	3	4	5	6	7	8	9	10	11	12
Cipher	N	I	I	G	X	G	A	M	X	C	A	D
Plain	.	I	L	.	.	.	.	E	O	.	.	R

The word after ARMY is probably WILL. This leads to the insertion of the letter W in the plain component and G in the cipher component. In a short time both components can be completely established.

g. In passing, it may be well to note that in the illustrative message in paragraph 50*a* the very frequent occurrence of tripled letters (MMM, WWW, FFF, etc.) indicates the presence of a frequently used short word, a frequently used ending, or the like, the letters of which are sequent in the plain component. An astute cryptanalyst who has noted the frequency of occurrence of such triplets could assume the value THE for them, go through the entire text replacing all triplets by THE, and then, by applying the principles of indirect symmetry, build up the plain component in a short time. With that much as a start, solution of the entire message would be considerably simplified.

h. The principles elucidated in this paragraph may, of course, also be applied to cases of progressive systems in which the progression is by intervals greater than 1, and, with necessary modifications, to cases in which the progression is not regular but follows a specific pattern, such as 1–2–3, 1–2–3, . . . , or 2–5–7–3–1, 2–5–7–3–1, and so. The latter types of progression are encountered in certain mechanical cryptographs, the study of which will be reserved for future texts.

SECTION XIV

THE "MONOALPHABETICITY" OR "Φ TEST"

52. Purpose of the Φ (phi) test.—*a.* The student has noted that the χ test is based upon the general theory of coincidences and employs the probability constants κ_p and κ_r. There is one more test of a related nature which may be useful for him to understand and its explanation will be given in the succeeding paragraphs.

b. In paragraph 48*e* it was stated that two monoalphabetic distributions when correctly combined will yield a single distribution which should still be monoalphabetic in character. This question arises, therefore, in the student's mind: Is there a test whereby he can ascertain methematically whether a distribution is monoalphabetic or not, especially in the case of one which has relatively few data? Such a test has been devised and is termed the "Φ (phi) test."

53. Derivation of the Φ test.—*a.* Consider a monographic or uniliteral frequency distribution which is monoalphabetic in composition. If there is a total of N letters in the distribution, in a system in which there are n possible elements, then there is a possible total of $\frac{N(N-1)}{2}$ pairs of letters (for comparison purposes).

b. Let the symbol f_A represent the number of occurrences of A, f_B the number of occurrences of B, and so on to f_Z. With regard to the letter A then, there are $\frac{f_A(f_A-1)}{2}$ coincidences. (Again the combinations of f_A things taken two at a time.) With regard to the letter B, there are $\frac{f_B(f_B-1)}{2}$ coincidences, and so on up to $\frac{f_Z(f_Z-1)}{2}$ coincidences for the letter Z. Now it has been seen that according to the κ test, in $\frac{N(N-1)}{2}$ comparisons of letters forming the two members of pairs of letters in normal English plain text, there should be $\frac{\kappa_p N(N-1)}{2}$ coincidences, where κ_p is the probability of monographic coincidence for the language in question.

c. Now the expected value of $\frac{f_A(f_A-1)}{2}+\frac{f_B(f_B-1)}{2}+\cdots+\frac{f_Z(f_Z-1)}{2}$ is equal to the theoretical number of coincidences to be expected in $\frac{N(N-1)}{2}$ comparisons of two letters, which for normal plain text is κ_p times $\frac{N(N-1)}{2}$ and for random text is κ_r times $\frac{N(N-1)}{2}$. That is, for plain text:

Expected value of $\frac{f_A(f_A-1)}{2}+\frac{f_B(f_B-1)}{2}+\cdots+\frac{f_Z(f_Z-1)}{2}=\kappa_p\times\frac{N(N-1)}{2}$, or

(IX) Expected value of $f_A(f_A-1)+f_B(f_B-1)+\ldots+f_Z(f_Z-1)=\kappa_p N(N-1)$; and for random text:

Expected value of $\frac{f_A(f_A-1)}{2}+\frac{f_B(f_B-1)}{2}+\cdots+\frac{f_Z(f_Z-1)}{2}=\kappa_r\times\frac{N(N-1)}{2}$, or

(X) Expected value of $f_A(f_A-1)+f_B(f_B-1)+\ldots+f_Z(f_Z-1)=\kappa_r N(N-1)$.
If for the left-hand side of equations (IX) and (X) the symbol $E(\Phi)$ is used, then these equations become:

(XI) For plain text . . . $E(\Phi_p)=\kappa_p N(N-1)$

(XII) For random text . . . $E(\Phi_r)=\kappa_r N(N-1)$,

where $E(\Phi)$ means the average or expected value of the expression in the parenthesis, κ_p and κ_r are the probabilities of monographic coincidence in plain and in random text, respectively.

d. Now in normal English plain text it has been found that $\kappa_p=.0667$. For random text of a 26-letter alphabet $\kappa_r=.038$. Therefore, equations (XI) and (XII) may now be written thus:

(XIII) For normal English plain text . . . $E(\Phi_p)=.0667\ N(N-1)$

(XIV) For random text (26-letter alphabet) . . . $E(\Phi_r)=.0385\ N(N-1)$

e. By employing equations (XIII) and (XIV) it becomes possible, therefore, to test a piece of text for monoalphabeticity or for "randomness." That is, by using these equations one can mathematically test a very short cryptogram to ascertain whether it is a monoalphabetically enciphered substitution or involves several alphabets so that for all practical purposes it is equivalent to random text. This test has been termed the Φ test.

54. Applying the Φ test.—*a*. Given the following short piece of text, is it likely that it is normal English plain text enciphered monoalphabetically?

A	B	C	D	E	F	G	H	I	J	K	L	M	N	O	P	Q	R	S	T	U	V	W	X	Y	Z
1			1	2	3	4	2			1				4	2		1						1		3

$N=25$

For this case the observed value of Φ is:

$$(1\times0)+(1\times0)+(2\times1)+(3\times2)+(4\times3)+(2\times1)+(1\times0)+(4\times3)+(2\times1)+(1\times0)+(1\times0)+(3\times2)=2+6+12+2+12+2+6=42$$

If this text were monoalphabetically enciphered English plain text the expected value of Φ is:

$$E(\Phi_p)=\kappa_p N(N-1)=.0667\times25\times24=40.0$$

If the text were random text, the expected value of Φ is:

$$E(\Phi_r)=\kappa_r N(N-1)=.0385\times25\times24=23.1$$

The conclusion is warranted, therefore, that the cryptogram is probably monoalphabetic substitution, since the observed value of Φ(42) more closely approximates the expected value for English plain text (40.0) than it does the expected value for random text (23.1). (As a matter of fact, the cryptogram was enciphered monoalphabetically.)

b. Here is another example. Given the following series of letters, does it represent a selection of English text enciphered monoalphabetically or does it more nearly represent a random selection of letters?

Y O U I J Z M M Z Z M R N Q C X I Y T W R G K L H

The distribution and calculation are as follows:

	A	B	C	D	E	F	G	H	I	J	K	L	M	N	O	P	Q	R	S	T	U	V	W	X	Y	Z
			/				/	/	//	/	/	/	///	/	/		/	//		/	/		/	/	//	///
$f(f-1)$ . . .			0	0	0	2	0	0	0	6	0	0		0	2		0	0		0	0	2	6			

$\Sigma f(f-1)=18$ (That is, observed value of $\Phi=18$)

$E(\Phi_p)=.0667\times 25\times 24=40.0$ (That is, expected value of $\Phi_p=40.0$)

The conclusion is that the series of letters does not represent a selection of English text monoalphabetically enciphered. Whether or not it represents a random selection of letters cannot be told, but it may be said that if the letters actually do constitute a cryptogram, the latter is probably polyalphabetically enciphered. (As a matter of fact, the latter statement is true, for the message was enciphered by 25 alphabets used in sequence.)

c. The Φ test is, of course, closely related to the χ test and derives from the same general theory as the latter, which is that of coincidence. When two monoalphabetic distributions have been combined into a single distribution, the Φ test may be applied to the latter as a check upon the χ test. It is also useful in testing the columns of a superimposition diagram, to ascertain whether or not the columns are monoalphabetic.

SECTION XV

CONCLUDING REMARKS

55. Concluding remarks on aperiodic substitution systems.—*a.* The various systems described in the foregoing pages represent some of the more common and well-known methods of introducing complexities in the general scheme of cryptographic substitution with the view to avoiding or suppressing periodicity. There are, of course, other methods for accomplishing this purpose, which, while perhaps a bit more complex from a practical point of view, yield more desirable results from a cryptographic point of view. That is, these methods go deeper into the heart of the problem of cryptographic security and thus make the task of the enemy cryptanalyst much harder. But studies based on these more advanced methods will have to be postponed at this time, and reserved for a later text.

b. Thus far in these studies, aside from a few remarks of a very general nature, no attention has been paid to that other large and important class of ciphers, viz, transposition. It is desirable, before going further with substitution methods, that the student gain some understanding of how to solve certain of the more simple varieties of transposition ciphers. Consequently, in the text to succeed the present text, the student will temporarily lay aside the various useful methods and tools that he has been given for the solution of substitution ciphers and will turn his thoughts toward the methods of breaking down transposition ciphers.

56. Synoptic table.—Continuing the plan instituted in previous texts, of summarizing the textual material in the form of a very condensed chart called An Analytical Key for Military Cryptanalysis, the outline for the studies covered by Part III is shown on p. 119.

APPENDIX 1

ADDITIONAL NOTES ON METHODS FOR SOLVING PLAIN-TEXT AUTO-KEYED CIPHERS

1. Introductory remarks.—*a.* In paragraph 33 of the text proper it was indicated that the method elucidated in paragraph 32 for solving plain-text auto-keyed ciphers is likely to be successful only if the cryptanalyst has been fortunate in his selection of a "probable word." Or, to put it another way, if the "probable words" which his imagination leads him to assume to be present in the text are really not present, then he is unfortunate, for solution will escape him. Hence, it is desirable to point out other principles and methods which are not so subject to chance. But because most of these methods are applicable only in special cases and because in general it is true that auto-key systems are no longer commonly encountered in practical military cryptography, it was thought best to exclude the exposition of these principles and methods from the text proper and to add them in an appendix, for the study of such students as find them of particular interest.

b. A complete discussion of the solution of plain-text auto-key systems, with examples, would require a volume in itself. Only one or two methods will be described, therefore, leaving the development of additional principles and methods to the ingenuity of the student who wishes to go more deeply into the subject. The discussion herein will be presented under separate headings, dependent upon the types of primary components employed.

c. As usual, the types of primary components may be classified as follows:

(1) Primary components are identical.

(a) Both components progress in the same direction.

(b) Both components progress in opposite directions.

(2) Primary components are different.

2. Simple "mechanical" solution.—*a.* (1) Taking up the case wherein the two identical primary components progress in the same direction, assume the following additional factors to be known by the cryptanalyst:

(a) The primary components are both normal sequences.

(b) The encipherment is by plain-text auto-keying.

(c) The enciphering equations are: $\Theta_{k/2}=\Theta_{i/1}$; $\Theta_{p/1}=\Theta_{c/2}$.

(2) A message beginning QVGLB TPJTF ... is intercepted; the only unknown factor is the initial key letter. Of course, one could try to decipher the message using each key letter in turn, beginning with A and continuing until the correct key letter is tried, whereupon plain text will be obtained. But it seems logical to think that all the 26 possible "decipherments" might be derived from the first one, so that the process might be much simplified, and this is true, as

will now be shown. Taking the two cipher groups under consideration, let them be "deciphered" with initial key letter A:

Cipher-- QVGLBTPJTF
Deciphered with keyletter A------------------------------------ QFBKRCNWXI

The deciphered text is certainly not "plain text." But if one completes the sequences initiated by these letters, using the direct standard sequence for the even columns, the reversed standard for the odd columns, the plain text sequence is seen to reappear on one generatrix: It is HOSTILE FOR(CE). From this it appears that instead of going through the labor of making 26 successive trials, which would consume considerable time, all that is necessary is to have a set of strips bearing the normal direct sequence and another set bearing the reversed normal sequence, and to align the strips, alternately direct and reversed, to the first "decipherment." The plain text will now reappear on one generatrix of the completion diagram. (See Fig. 1.)

Initial key letter	Q	V	G	L	B	T	P	J	T	F	
A	Q	F	B	K	R	C	N	W	X	I	
B	P	G	A	L	Q	D	M	X	W	J	
C	O	H	Z	M	P	E	L	Y	V	K	
D	N	I	Y	N	O	F	K	Z	U	L	
E	M	J	X	O	N	G	J	A	T	M	
F	L	K	W	P	M	H	I	B	S	N	
G	K	L	V	Q	L	I	H	C	R	O	
H	J	M	U	R	K	J	G	D	Q	P	
I	I	N	T	S	J	K	F	E	P	Q	
J	H	O	S	T	I	L	E	F	O	R	*
K	G	P	R	U	H	M	D	G	N	S	
L	F	Q	Q	V	G	N	C	H	M	T	
M	E	R	P	W	F	O	B	I	L	U	
N	D	S	O	X	E	P	A	J	K	V	
O	C	T	N	Y	D	Q	Z	K	J	W	
P	B	U	M	Z	C	R	Y	L	I	X	
Q	A	V	L	A	B	S	X	M	H	Y	
R	Z	W	K	B	A	T	W	N	G	Z	
S	Y	X	J	C	Z	U	V	O	F	A	
T	X	Y	I	D	Y	V	U	P	E	B	
U	W	Z	H	E	X	W	T	Q	D	C	
V	V	A	G	F	W	X	S	R	C	D	
W	U	B	F	G	V	Y	R	S	B	E	
X	T	C	E	A	U	Z	Q	T	A	F	
Y	S	D	D	I	T	A	P	U	Z	G	
Z	R	E	C	J	S	B	O	V	Y	H	

FIGURE 1.

b. The peculiar nature of the phenomenon just observed, *viz*, a completion diagram with the vertical sequences in adjacent columns progressing in opposite directions, those in alternate columns in the same direction, calls for an explanation. Although the matter seems rather mysterious, it will not be hard to understand. First, it is not hard to see why the letters in column 1 of Figure 1 should form the descending sequence QPO. . . for these letters are merely the ones resulting from the successive "decipherment" of Q_c by the successive key letters A, B, C, ... Now since the "decipherment" obtained from the 1st cipher letter in any row in Figure 1 becomes the key letter for "deciphering" the 2d cipher letter in the same row, it is apparent that as the letters in the 1st column progress in a reversed normal (descending) order, the letters in the 2d column *must* progress in a direct normal (ascending) order. The matter may perhaps become more clear if encipherment is regarded as a process of addition and decipherment as a process of subtraction. Instead of primary components or a Vigenère square, one may use simple arithmetic, assigning numerical values to the letters of the alphabet, beginning with A=Ø and ending with Z=25. Thus on the basis of the pair of enciphering equations $\Theta_{k/2}=\Theta_{1/1}$; $\Theta_{p/1}=\Theta_{c/2}$, the letter H_c enciphered by key letter M_k with direct primary components yields T_c. But using the following numerical values:

A	B	C	D	E	F	G	H	I	J	K	L	M	N	O	P	Q	R	S	T	U	V	W	X	Y	Z
0	1	2	3	4	5	6	7	8	9	10	11	12	13	14	15	16	17	18	19.	20	21	22	23	24	25

the same result may be obtained thus: $H_p(M_k)=7+12=19=T_c$. Every time the number 25 is exceeded in the addition, one subtracts 26 from it and finds the letter equivalent for the remainder. In decipherment, the process is one of subtraction.[1] For example: $T_c(M_k=19-12=7=H_p$; $D_c(R_k)=3-17=[(26+3)-17]=29-17=12=M_p$. Using this arithmetical equivalent of normal sliding-strip encipherment, the phenomenon just noted can be set down in the form of a diagram (Fig. 2) which will perhaps make the matter clear.

[1] It will be noted that if the letters of the alphabet are numbered from 1 to 26, in the usual manner, the arithmetical method must be modified in a minor particular in order to obtain the same results as are given by employing the normal Vigenère square. This modification consists merely in subtracting 1 from the numerical value of the key letter. Thus:

A	B	C	D	E	F	G	H	I	J	K	L	M	N	O	P	Q	R	S	T	U	V	W	X	Y	Z
1	2	3	4	5	6	7	8	9	10	11	12	13	14	15	16	17	18	19	20	21	22	23	24	25	26

$$H_p(M_k)=8+(13-1)=8+12=20=T_c$$
$$T_c(M_k)=20-(13-1)=20-12=8=H_p$$

For an interesting extension of the basic idea involved in arithmetic cryptography, see:

Hill, Lester S. *Cryptography in an Algebraic Alphabet.* American Mathematical Monthly, Vol. XXXVI, No. 6, 1929.

Ibid. *Concerning certain linear transformation apparatus of cryptography.* American Mathematical Monthly, Vol. XXXVIII, No. 3, 1931.

Q V G L B T P J etc.

$Q_c(A_k)=16-0=16=Q$ → Q F B K R . . .
$V_c(Q_k)=21-16=5=F$
$G_c(F_k)=6-5=1=B$
$L_c(B_k)=11-1=10=K$
$B_c(K_k)=1-10=17=R$

* *

$Q_c(B_k)=16-1=15=P$ → P G A L Q . . .
$V_c(P_k)=21-15=6=G$
$G_c(G_k)=6-6=0=A$
$L_c(A_k)=11-0=11=L$
$B_c(L_k)=1-11=16=Q$

* *

$Q_c(C_c)=16-2=14=O$ → O H Z M P . . .
$V_c(O_k)=21-14=7=H$
$G_c(H_k)=6-7=25=Z$
$L_c(Z_k)=11-25=12=M$
$B_c(M_k)=1-12=15=P$

FIGURE 2.

Note how homologous letters of the three rows (joined by vertical dotted lines) form alternately descending and ascending normal sequences.

c. When the method of encipherment based upon enciphering equations $\theta_{k/2}=\theta_{i/1}$; $\theta_{p/2}=\theta_{c/1}$ is used instead of the one based upon enciphering equations $\theta_{k/2}=\theta_{i/1}$; $\theta_{p/1}=\theta_{c/2}$, the process indicated above is simplified by the fact that no alternation in the direction of the sequences in the completion diagram is required. For example:

```
Cipher.................. Y H E B P D T B J D
Deciphered A=A.......... Y F J K Z C V W F I
                         -------------------
                         Z G K L A D W X G J
                         A H L M B E X Y H K
                         B I M N C F Y Z I L
                         C J N O D G Z A J M
                         D K O P E H A B K N
                         E L P Q F I B C L O
                         F M Q R G J C D M P
                         G N R S H K D E N Q
                        *H O S T I L E F O R
```

FIGURE 3.

d. (1) In the foregoing example the primary components were normal sequences, but the case of identical mixed components may be handled in a similar manner. Note the following example, based upon the following primary component (which is assumed to have been reconstructed from previous work):

F B P Y R C Q Z I G S E H T D J U M K V A L W N O X

Message.............. U S I N L Y Q E O P . . . etc.

(2) First, the message is "deciphered" with the initial key-letter A, and then a completion diagram is established, using sliding strips bearing the mixed primary component, alternate strips bearing the reversed sequence. Note Figure 4, in which the plain text, HOSTILE FOR(CE), reappears on a single generatrix. Note also that whereas in Figure 1 the odd columns contain the primary sequence in the reversed order, and the even columns contain the sequence in the direct order, in Figure 4 the situation is reversed: the odd columns contain the primary sequence in the direct order, and the even columns contain the sequence in the reversed order. This point is brought to notice to show that it is immaterial whether the direct order is used for odd columns or for even columns; the *alternation in direction* is all that is required in this type of solution.

e. (1) There is next to be considered the case in which the two primary components progress in opposite directions [par. 1*c* (1) (b)]. Here is a message, known to have been enciphered by reversed standard alphabets, plain-text auto-keying having been followed:

X T W Z L X H Z R X

(2) The procedure in this case is exactly the same as before, except that it is not necessary to have any alternation in direction of the completion sequences, which may be either that of the plain component or the cipher component. Note the solution in Figure 5. Let the student ascertain why the alternation in direction of the completion sequences is not necessary in this case.

(3) In the foregoing case the alphabets were reversed standard, produced by the sliding of the normal sequence against its reverse. But the underlying principle of solution is the same even if a mixed sequence were used instead of the normal; so long as the sequence is known, the procedure to be followed is exactly the same as demonstrated in subparagraphs (1) and (2) hereof. Note the following solution:

MESSAGE

V D D N C T S E P A . . .

Plain component------- F B P Y R C Q Z I G S E H T D J U M K V A L W N O X
Cipher component----- X O N W L A V K M U J D T H E S G I Z Q C R Y P B F

Note here that the primary mixed sequence is used for the completion sequence and that the plain text, HOSTILE FOR(CE), comes out on one generatrix. It is immaterial whether the direct or reversed mixed component is used for the completion sequence, so long as *all* the sequences in the diagram progress in the same direction. (See Fig. 6.)

f. (1) There remains now to be considered only the case in which the two components are different mixed sequences. Let the two primary components be as follows:

Plain---------------- A B C D E F G H I J K L M N O P Q R S T U V W X Y Z
Cipher-------------- F B P Y R C Q Z I G S E H T D J U M K V A L W N O X

and the message:

C F U Y L V X U D J

```
U S I N L Y Q E O P
-------------------
W D A Y K E L U I A
N T L P V S W J G V
O H W B A G N D S K
X E N F L I O T E M
F S O X W Z X H H U
B G X O N Q F E T J
P I E N O C B S D D
Y Z B W X R P G J T
R Q P L F Y Y I U H
C C Y A B P R Z M E
Q R R V P B C Q K S
Z Y C K Y F Q C V G
I P Q M R X Z R A I
G B Z U C O I Y L Z
S F I J Q N G P W Q
E X G D Z W S B N C
H O S T I L E F O R*
T N E H G A H X X Y
D W H E S V T O F P
J L T S E K D N B P
U A D G H M J W P F
M V J I T U U L Y X
K K U Z D J M A R O
V M M Q I D K V C N
A U K C U T V K Q W
L J V R M H A M Z L
↓ ↑ ↓ ↑ ↓ ↑ ↓ ↑ ↓ ↑
```

FIGURE 4.

```
X T W Z L X H Z R X
C J N O D G Z A J M
-------------------
D K O P E H A B K N
E L P Q F I B C L O
F M Q R G J C D M P
G N R S H K D E N Q
H O S T I L E F O R*
I P T U J M F G P S
J Q U V K N G H Q T
K R V W L O H I R U
L S W X M P I J S V
M T X Y N Q J K T W
N U Y Z O R K L U X
O V Z A P S L M V Y
P W A B Q T M N W Z
Q X B C R U N O X A
R Y C D S V O P Y B
S Z D E T W P Q Z C
T A E F U X Q R A D
U B F G V Y R S B E
V C G H W Z S T C F
W D H I X A T U D G
X E I J Y B U V E H
Y F J K Z C V W F I
Z G K L A D W X G J
A H L M B E X Y H K
B I M N C F Y Z I L
```

FIGURE 5.

```
V D D N C T S E P A
Z V C I Y U Q L V X
-------------------
I A Q G R M Z W A F
G L Z S C K I N L B
S W I E Q V G O W P
E N G H Z A S X N Y
H O S T I L E F O R*
T X E D G W H B X C
D F H J S N T P F Q
J B T U E O D Y B Z
U P D M H X J R P I
M Y J K T F U C Y G
K R U V D B M Q R S
V C M A J P K Z C E
A Q K L U Y V I Q H
L Z V W M R A G Z T
W I A N K C L S I D
N G L O V Q W E G J
O S W X A Z N H S U
X E N F L I O T E M
F H O B W G X D H K
B T X P N S F J T V
P D F Y O E B U D A
Y J B R X H P M J L
R U P C F T Y K U W
C M Y Q B D R V M N
Q K R Z P J C A K O
Z V C I Y U Q L V X
```

FIGURE 6.

(2) First "decipher" the message with any arbitrarily selected initial key letter, say **A**, and complete the plain component sequence in the first column (Fig. 7*a*).

Cipher......	C	F	U	Y	L	V	X	U	D	J
Plain........	L	F	Q	X	W	X	A	W	S	F
	M									
	N									
	O									
	P									
	Q									
	R									
	S									
	T									
	U									
	V									
	W									
	X									
	Y									
	Z									
	A									
	B									
	C									
	D									
	E									
	F									
	G									
	H									
	I									
	J									
	K									

FIGURE 7*a*.

C	F	U	Y	L	V	X	U	D	J
L	F	Q	X	W	X	A	W	S	E
M	J								
N	D								
O	C								
P	Y								
Q	U								
R	W								
S	Q								
T	N								
U	K								
V	H								
W	E								
X	B								
Y	X								
Z	T								
A	G								
B	Z								
C	V								
D	M								
E	P								
F	A								
G	R								
H	O								
I	S								
J	L								
K	I								

FIGURE 7*b*.

C	F	U	Y	L	V	X	U	D	J
L	F	Q	X	W	X	A	W	S	E
M	J	B	C						
N	D	C	Y						
O	C	L	I						
P	Y	N	G						
Q	U	A	J						
R	W	U	N						
S	Q	K	L						
T	N	T	Q						
U	K	Y	A						
V	H	E	S						
W	E	F	D						
X	B	P	B						
Y	X	R	Z						
Z	T	D	P						
A	E	H	R						
B	Z	J	O						
C	V	X	E						
D	M	Z	W						
E	P	O	F						
F	A	W	H						
G	R	M	M						
*H	O	S	T						
I	S	G							
J	L	V							
K	I	I							

FIGURE 7*c*.

Now prepare a strip bearing the cipher component *reversed*, and set it below the plain component so that $F_p=L_c$, a setting given by the 1st two letters of the spurious "plain text" recovered. Thus:

Plain..........	A	B	C	D	E	F	G	H	I	J	K	L	M	N	O	P	Q	R	S	T	U	V	W	X	Y	Z
Cipher..........	F	X	O	N	W	L	A	V	K	M	U	J	D	T	H	E	S	G	I	Z	Q	C	R	Y	P	B

(3) Now opposite each letter of the completion sequence in column 1, write its plain-component equivalent, as given by the juxtaposed sequences above. This gives what is shown in Figure 7*b*. Then reset the two sequences (reversed cipher component and the plain component) so that $Q_p=F_c$ (to correspond with the 2d and 3d letters of the spurious plain text); write down the plain-component equivalents of the letters in column 2, forming column 3. Continue this process, scanning the generatrices from time to time, resetting the two components and finding equivalents from column to column, until it becomes evident on what generatrix the plain text is reappearing. In Figure 7*c* it is seen that the plain text generatrix is the one beginning HOST, and from this point on the solution may be obtained directly, by using the two primary components.

(4) When the plain component is also a mixed sequence (and different from the cipher component), the procedure is identical with that outlined in subparagraphs (1)–(3) above. The fact that the plain component in the preceding case is the normal sequence is of no particular significance in the solution, for it acts as a mixed sequence would act under similar circumstances. To demonstrate, suppose the two following components were used in encipherment of the message below:

```
Plain................ W B V I G X L H Y A J Z M N F O R P E Q D S C T K U
Cipher............... F B P Y R C Q Z I G S E H T D J U M K V A L W N O X
```

```
Message.......... B B V Z U   D Q X J D ...
```

To solve the message, "decipher" the text with any arbitrarily selected initial key letter and proceed exactly as in subparagraphs (2) and (3) above. Thus:

```
Cipher........................ B B V Z U D Q X J D
"Plain" (Θk=X)................ V Y R I Y Z E F O R
```

Note the completion diagram in Figure 8 which shows the word HOST... very soon in the process. From this point on the solution may be obtained directly, by using the two primary components.

```
 B B V Z U D Q X J D
 V Y R I Y Z E F O R
 -------------------
 I Q N J
 G E Y G
 X V W Z
 L L K O
*H O S T
 Y K B
 A H H
 J M V
 Z D X
 M J G
 N G J
 F B E
 O I Z
 R T L
 P U I
 E R O
 Q S A
 D N C
 S P P
 C C F
 T F Q
 K A U
 U Z M
 W X D
```

FIGURE 8.

3. Another "mechanical" solution.—*a.* Another "mechanical" solution for the foregoing cases will now be described because it presents rather interesting cryptanalytic sidelights. Take the message

R E F E R E N C E   H I S   P R E F E R E N C E   I N   R E F E R E N C E
B O O K S   A N D   R E F E R E N C E   C H A R T S . . .

and encipher it by plain-text auto-key, with normal direct primary components, initial key setting $A_p=G_c$. Then note the underscored repetitions:

R E F E R E N C E H I S P R E F E R E N C E I N R E F E
X V J J V V R P G L P A H G V J J V V R P G M V E V J J

R E N C E B O O K S A N D R E F E R E N C E C H A R T S
V V R P G F P C Y C S N Q U V J J V V R P G G J H Y K L

b. Now suppose the message has been intercepted and is to be solved. The only unknown factor will be assumed to be the initial key letter. Let the message be "deciphered" by means of any initial key letter,[2] say A, and then note the underscored repetitions in the spurious plain text.

Cipher............. X V J J V V R P G L P A H G V J J V V R P G M V E V J J
"Plain text"...... X Y L Y X Y T W K B O M V L K Z K L K H I Y O H X Y L Y

Cipher............. V V R P G F P C Y C S N Q U V J J V V R P G G J H Y K L
"Plain text"...... X Y T W K V U I Q M G H J L K Z K L K H I Y I B G S S T

The original four 8-letter repetitions now turn out to be two different sets of 9-letter repetitions. This calls for an explanation. Let the spurious plain text, with its real plain text be transcribed as though one were dealing with a periodic cipher involving two alphabets, as shown in Figure 9. It will here be seen that the letters in column 1 are monoalphabetic, and so are those in column 2. In other words, an auto-key cipher, which is commonly regarded as a polyalphabetic, aperiodic cipher, has been converted into a 2-alphabet, periodic cipher, the individual alphabets of which are now monoalphabetic in nature. The two repetitions of X Y L Y X Y T W K represent encipherments of the word REFERENCE, in alphabets 1-2-1-2-1-2-1-2-1; the two repetitions of L K Z K L K H I Y likewise represent encipherments of the same word but in alphabets 2-1-2-1-2-1-2-1-2.

c. Later on it will be seen how this method of converting an auto-key cipher into a periodic cipher may be applied to the case where an introductory key word is used as the initial keying element instead of a single letter, as in the present case.

1–2	1–2	1–2	1–2
R E	E F	R E	E F
X Y	K Z	X Y	K Z
F E	E R	N C	E R
L Y	K L	T W	K L
R E	E N	E B	E N
X Y	K H	K V	K H
N C	C E	O O	C E
T W	I Y	U I	I Y
E H	I N	K S	C H
K B	O H	Q M	I B
I S	R E	A N	A R
O M	X Y	G H	G S
P R	F E	D R	T S
V L	L Y	J L	S T

FIGURE 9.

[2] Except the actual key letter or a letter 13 intervals from it. See subparagraph (7) below.

d. The student has probably already noted that the phenomena observed in this subparagraph are the same as those observed in subparagraph 2*b*. In the latter subparagraph it was seen that the direction of the sequences in alternate columns had to be reversed in order to bring out the plain text on one generatrix. If this reversal is not done, then obviously the plain text would appear on *two* generatrices, which is equivalent to having the plain text reduced to two monoalphabets.

e. When reciprocal components are employed, the spurious plain text obtained by "decipherment" with a key setting other than the actual one will be monoalphabetic throughout. Note the following encipherment (with initial key setting $A_p=G_c$, using a reversed standard sequence sliding against the direct standard) and its "decipherment" by setting these two components $A_p=A_c$.

Plain text	R E F E R E N C E H I S P R E F E R E N C E . . .
Cipher	P N Z B N N R L Y X Z Q D Y N Z B N N R L Y . . .
Spurious plain text	L Y Z Y L Y H W Y B C M J L Y Z Y L Y H W Y . . .

Here the spurious plain text is wholly monoalphabetic.

f. The reason for the exception noted in footnote 2 on page 106 now becomes clear. For if the actual initial key letter (G) were used, of course the decipherment yields the correct plain text; if a letter 13 intervals removed from G is used as the key letter, the cipher alphabet selected for the first "decipherment" is the reciprocal of the real initial cipher alphabet and thereafter all alternate cipher alphabets are reciprocal. Hence the spurious text obtained from such a "decipherment" must be monoalphabetic.

g. In the foregoing case the primary components were identical normal sequences progressing in the same direction. If they were mixed sequences the phenomena observed above would still hold true, and so long as the sequences are known, the indicated method of solution may be applied.

h. When the two primary components are known but differently mixed sequences, this method of solution is too involved to be practical. It is more practicable to try successive initial key letters, noting the plain text each time and resetting the strips until the correct setting has been ascertained, as will be evidenced by obtaining intelligible plain text.

4. Solution of plain-text auto-keyed cryptograms when the introductory key is a word or phrase.—*a.* In the foregoing discussion of plain-text auto-keying, the introductory key was assumed to consist of a single letter, so that the subsequent key letters are displaced one letter to the right with respect to the text of the message itself. But sometimes a word or phrase may serve this function, in which case the subsequent key is displaced as many letters to the right of the initial plain-text letter of the message as there are letters in the initial key. This will not, as a rule, interfere in any way with the application of the principles of solution set forth in paragraph 28 to that part of the cryptogram subsequent to the introductory key, and a solution by the probable-word method and the study of repetitions can be reached. However, it may happen that trial of this method is not successful in certain cryptograms because of the paucity of repetitions, or because of failure to find a probable word in the text. When the cipher alphabets are known there is another point of attack which is useful and interesting. The method consists in finding the length of the introductory key and then solving by frequency principles. Just how this is accomplished will now be explained.

b. Suppose that the introductory key word is HORSECHESTNUT, that the plain-text message is as below, and that identical primary components progressing in the same direction are used to encipher the message, by enciphering equation $\Theta_{k/2}=\Theta_{1/1}$; $\Theta_{p/1}=\Theta_{c/2}$. Let the components be the normal sequence. The encipherment is as follows:

	1	2	3	4	5	6	7	8	9	10	11	12	13	14	15	16	17	18	19	20	21	22	23	24	25	26
Key	H	O	R	S	E	C	H	E	S	T	N	U	T	M	Y	L	E	F	T	F	L	A	N	K	I	S
Plain	M	Y	L	E	F	T	F	L	A	N	K	I	S	R	E	C	E	I	V	I	N	G	H	E	A	V
Cipher	T	M	C	W	J	V	M	P	S	G	X	C	L	D	C	N	I	N	O	N	Y	G	U	O	I	N
Key	R	E	C	E	I	V	I	N	G	H	E	A	V	Y	A	R	T	I	L	L	E	R	Y	F	I	R
Plain	Y	A	R	T	I	L	L	E	R	Y	F	I	R	E	E	N	E	M	Y	I	S	M	A	S	S	I
Cipher	P	E	T	X	Q	G	T	R	X	F	J	I	M	C	E	E	X	U	J	T	W	D	Y	X	A	Z
Key	E	E	N	E	M	Y	I	S	M	A	S	S	I	N	G	T	R	O	O	P	S	T	O	L	E	F
Plain	N	G	T	R	O	O	P	S	T	O	L	E	F	T	F	R	O	N	T	A	N	D	C	O	N	C
Cipher	R	K	G	V	A	M	X	K	F	O	D	W	N	G	L	K	F	B	H	P	F	W	Q	Z	R	H
Key	T	F	R	O	N	T	A	N	D	C	O	N	C	E	N	T	R	A	T	I	N	G	A	R	T	I
Plain	E	N	T	R	A	T	I	N	G	A	R	T	I	L	L	E	R	Y	T	H	E	R	E	X	W	I
Cipher	X	S	K	F	N	M	I	A	J	C	F	G	K	P	Y	X	I	Y	M	P	R	X	E	O	P	Q
Key	L	L	E	R	Y	T	H	E	R	E	X	W	I	L	L	N	E	E	D	C	O	N	S	I	D	E
Plain	L	L	N	E	E	D	C	O	N	S	I	D	E	R	A	B	L	E	R	E	I	N	F	O	R	C
Cipher	W	W	R	V	C	W	J	S	E	W	F	Z	M	C	L	O	P	I	U	G	W	A	X	W	U	G
Key	R	A	B	L	E	R	E	I	N	F	O	R	C	E	M	E	N	T	S	T	O	M	A	I	N	T
Plain	E	M	E	N	T	S	T	O	M	A	I	N	T	A	I	N	M	Y	P	O	S	I	T	I	O	N
Cipher	V	M	F	Y	X	J	X	W	Z	F	W	E	V	E	U	R	Z	R	H	H	G	U	T	Q	B	G

It will now be noted that since the introductory key contains 13 letters the 14th letter of the message is enciphered by the 1st letter of the plain text, the 15th by the 2d, and so on. Likewise, the 27th letter is enciphered by the 14th, the 28th by the 15th, and so on. Hence, if the 1st cipher letter is deciphered, this will give the key for deciphering the 14th, the latter will give the key for the 27th, and so on. An important step in the solution of a message of this kind would therefore involve ascertaining the length of the introductory key. This step will now be explained.

c. Since the plain text itself constitutes the key letters in this system (after the introductory key), these key letters will occur with their normal frequencies, and this means that there will be many occurrences of E, T, O, A, N, I, R, S, enciphered by E_k; there will be many occurrences of these same high-frequency letters enciphered by T_k, by O_k, by A_k, and so on. In fact, the number of times each of these combinations will occur may be calculated statistically. With the enciphering conditions set forth under *b* above, E_p enciphered by T_k, for example, will yield the same cipher equivalent as T_p enciphered by E_k; in other words two encipherments of any pair of letters of which either may serve as the key for enciphering the other must yield the same cipher resultant.[3] It is the cryptographic effect of these two phenomena working together which permits of ascertaining the length of the introductory key in such a case. For every time a given letter, Θ_p, occurs in the plain text it will occur *n* letters later as a key letter, Θ_k, and *n* in this case equals the length of the introductory key. Note the following illustration:

[3] It is important to note that the two components must be identical sequences and progress in the same direction. If this is not the case, the entire reasoning is inapplicable.

	1	2	3	4	5	6	7	8	9	10	11	12	13	1	2	3	4	5	6	7	8	9	10	11	12	13
(1) Key	H	O	R	S	E	C	H	E	S	T	N	U	T	.	.	.	.	.	T	.	.	.	.	.	.	.
(2) Plain	.	.	.	.	.	T	.	.	.	.	.	.	.	.	.	.	.	.	E	.	.	.	.	.	.	.
(3) Cipher	.	.	.	.	.	.	.	.	.	.	.	.	.	.	.	.	.	.	X	.	.	.	.	.	.	.

	1	2	3	4	5	6	7	8	9
(1) Key	.	.	.	.	.	E	.	.	.
(2) Plain	.	.	.	.	.	T	.	.	.
(3) Cipher	.	.	.	.	.	X	.	.	.

Here it will be noted that E_p in line (2) has a T_p on either side of it, at a distance of 13 intervals; the first encipherment (E_p by T_k) yields the same equivalent (X_c) as the second encipherment (T_p by E_k). Two cipher letters are here identical, at an interval equal to the length of the introductory key. But the converse is not true; that is, not *every* pair of identical letters in the cipher text represents a case of this type. For in this system identity in two cipher letters may be the result of the following three conditions each having a statistically ascertainable probability of occurrence:

(1) A given plain-text letter is enciphered by the same key letter two different times, at an interval which is purely accidental; the cipher equivalents are identical but could not be used to give any information about the length of the introductory key.

(2) Two different plain-text letters are enciphered by two different key letters; the cipher equivalents are fortuitously identical.

(3) A given plain-text letter is enciphered by a given key letter and later on the same plain-text letter serves to encipher another plain-text letter which is identical with the first key letter; the cipher equivalents are causally identical.

It can be proved that the probability for identities of the third type is greater than that for identities of either or both 1st and 2d types *for that interval which corresponds with the length of the introductory key*; that is, if a tabulation is made of the intervals between identical letters in such a system as the one being studied, the interval which occurs most frequently should coincide with the length of the introductory key. The demonstration of the mathematical basis for this fact is beyond the scope of the present text; but a practical demonstration will be convincing.

d. Let the illustrative message be transcribed in lines of say 11, 12, and 13 letters, as in Figure 10.

1	2	3	4	5	6	7	8	9	10	11
T	M	C	W	J	V	M	P	S	G	X
C	L	D	C	N	I	N	O	N	Y	G
U	O	I	N	P	E	T	X	Q	G	T
R	X	F	J	I	M	C	E	E	X	U
J	T	W	D	Y	X	A	Z	R	K	G
V	A	M	X	K	F	O	D	W	N	G
L	K	F	B	H	P	F	W	Q	Z	R
H	X	S	K	F	N	M	I	A	J	C
F	G	K	P	Y	X	I	Y	M	P	R
X	E	O	P	Q	W	W	R	V	C	W
J	S	E	W	F	Z	M	C	L	O	P
I	U	G	W	A	X	W	U	G	V	M
F	Y	X	J	X	W	Z	F	W	E	V
E	U	R	Z	R	H	H	G	U	T	Q
B	G									

1	2	3	4	5	6	7	8	9	10	11	12
T	M	C	W	J	V	M	P	S	G	X	C
L	D	C	N	I	N	O	N	Y	G	U	O
I	N	P	E	T	X	Q	G	T	R	X	F
J	I	M	C	E	E	X	U	J	T	W	D
Y	X	A	Z	R	K	G	V	A	M	X	K
F	O	D	W	N	G	L	K	F	B	H	P
F	W	Q	Z	R	H	X	S	K	F	N	M
I	A	J	C	F	G	K	P	Y	X	I	Y
M	P	R	X	E	O	P	Q	W	W	R	V
C	W	J	S	E	W	F	Z	M	C	L	O
P	I	U	G	W	A	X	W	U	G	V	M
F	Y	X	J	X	W	Z	F	W	E	V	E
U	R	Z	R	H	H	G	U	T	Q	B	G

1	2	3	4	5	6	7	8	9	10	11	12	13
T	M	C	W	J	V	M	P	S	G	X	C	L
D	C	N	I	N	O	N	Y	G	U	O	I	N
P	E	T	X	Q	G	T	R	X	F	J	I	M
C	E	E	X	U	J	T	W	D	Y	X	A	Z
R	K	G	V	A	M	X	K	F	O	D	W	N
G	L	K	F	B	H	P	F	W	Q	Z	R	H
X	S	K	F	N	M	I	A	J	C	F	G	K
P	Y	X	I	Y	M	P	R	X	E	O	P	Q
W	W	R	V	C	W	J	S	E	W	F	Z	M
C	L	O	P	I	U	G	W	A	X	W	U	G
V	M	F	Y	X	J	X	W	Z	F	W	E	V
E	U	R	Z	R	H	H	G	U	T	Q	B	G

b

FIGURE 10.

462861 O - 42 - 8

In each transcription, every pair of superimposed letters is noted and the number of identities is indicated by ringing the letters involved, as shown above. The number of identities for an assumed introductory-key length 13 is 9, as against 3 for the assumption of a key of 11 letters, and 5 for the assumption of a key of 12 letters.

e. Once having found the length of the introductory key, two lines of attack are possible: the composition of the key may be studied, which will yield sufficient plain text to get a start toward solution; or, the message may be reduced to periodic terms and solved as a repeating-key cipher. The first line of attack will be discussed first, it being constantly borne in mind in this paragraph that the entire discussion is based upon the assumption that the cipher alphabets are known alphabets. The illustrative message of *b* above will be used.

5. Subsequent steps after determining the length of the introductory key.—*a.* Assume that the first letter of the introductory key is A and decipher the 1st cipher letter T_c (with direct standard alphabets). This yields T_p and the latter becomes the key letter for the 14th letter of the message. The 14th letter is deciphered: D_c $(T_k)=K_p$; the latter becomes the key letter for the 27th letter and so on, down the entire first column of the message as transcribed in lines of 13 letters. The same procedure is followed using B as the initial key letter, then C, and so on. The message as it appears for the first three trials (assuming A, B, then C as the initial key letter) is shown in Figure 11.

(a)	(b)	(c)
1 2 3 4 5 6 7 8 9 10 11 12 13	1 2 3 4 5 6 7 8 9 10 11 12 13	1 2 3 4 5 6 7 8 9 10 11 12 13
T M C W J V M P S G X C L	T M C W J V M P S G X C L	T M C W J V M P S G X C L
T	S	R
D	D	D
K	L	M
P	P	P
F	E	D
C	C	C
X	Y	Z
R	R	R
U	T	S
G	G	G
M	N	O
X	X	X
L	K	J
P	P	P
E	F	G
W	W	W
S	R	Q
C	C	C
K	L	M
V	V	V
L	K	J
E	E	E
T	U	V

(a) First column of Figure 10 (c) "deciphered" with initial θ_k=A. | (b) First column of Figure 10 (c) "deciphered" with initial θ_k=B. | (c) First column of Figure 10 (c) "deciphered" with θ_k=C.

FIGURE 11.

b. Inspection of the results of these three trials soon shows that the entire series of 26 trials need not be made, for the results can be obtained from the very first trial. This may be shown graphically by superimposing merely the results of the first three trials *horizontally*. Thus:

		T	D	P	C	R	G	X	P	W	C	V	E
Cipher letters of Col. 1, Fig. 11		T	D	P	C	R	G	X	P	W	C	V	E
Keyletters	A	T	K	F	X	U	M	L	E	S	K	L	T
	B	S	L	E	Y	T	N	K	F	R	L	K	U
	C	R	M	D	Z	S	O	J	G	Q	M	J	V
	D	↑	↓	↑	↓	↑	↓	↑	↓	↑	↓	↑	↓
	E												

FIGURE 12.

c. It will be noted that the vertical sequences in adjacent columns proceed in opposite directions, whereas those in alternate columns proceed in the same direction. The explanation for this alternation in progression is the same as in the previous case wherein this phenomenon was encountered (par. 2*b*), and the sequences in Figure 12 may now be completed very quickly. The diagram becomes as shown in Figure 13.

d. *One* of the horizontal lines or generatrices of figure 13 is the correct one; that is, it contains the actual plain-text equivalents of the 1st, 14th, 27th, . . . letters of the message. The correct generatrix can be selected by mere ocular examination, as is here possible (see generatrix marked by asterisk in Fig. 13), or it may be selected by a frequency test, assigning weights to each letter according to its normal plain-text frequency. (See par. 14*f* of *Military Cryptanalysis, Part II.*)

T	D	P	C	R	G	X	P	W	C	V	E
T	K	F	X	U	M	L	E	S	K	L	T
S	L	E	Y	T	N	K	F	R	L	K	U
R	M	D	Z	S	O	J	G	Q	M	J	V
Q	N	C	A	R	P	I	H	P	N	I	W
P	O	B	B	Q	Q	H	I	O	O	H	X
O	P	A	C	P	K	G	J	N	P	G	Y
N	Q	Z	D	O	S	F	K	M	Q	F	Z
M	R	Y	E	N	T	E	L	L	R	E	A*
L	S	X	F	M	U	D	M	K	S	D	B
K	T	W	G	L	V	C	N	J	T	C	C
J	U	V	H	K	W	B	O	I	U	B	D
I	V	U	I	J	X	A	P	H	V	A	E
H	W	T	J	I	Y	Z	Q	G	W	Z	F
G	X	S	K	H	Z	Y	R	F	X	Y	G
F	Y	R	L	G	A	X	S	E	Y	X	H
E	Z	Q	M	F	B	W	T	D	Z	W	I
D	A	P	N	E	C	V	U	C	A	V	J
C	B	O	O	D	D	U	V	B	B	U	K
B	C	N	P	C	E	T	W	A	C	T	L
A	D	M	Q	B	F	S	X	Z	D	S	M
Z	E	L	R	A	G	R	Y	Y	E	R	N
Y	F	K	S	Z	H	Q	Z	X	F	Q	O
X	G	J	T	Y	I	P	A	W	G	P	P
W	H	I	U	X	J	O	B	V	H	O	Q
V	I	H	V	W	K	N	C	U	I	N	R
U	J	G	W	V	L	M	D	T	J	M	S

FIGURE 13.

e. Identical procedure is followed with respect to columns 2, 3, 4, . . . of Figure 10*c*, with the result that the initial key word HORSECHESTNUT is reconstructed and the whole message may be now deciphered quite readily.

6. Conversion of foregoing aperiodic cipher into periodic form.—*a*. In paragraph 4 it was stated that an aperiodic cipher of the foregoing type may be reduced to periodic terms and solved as though it were a repeating-key cipher, provided the primary components are known sequences. The basis of the method lies in the phenomena noted in paragraph 2*b*. An example will be given.

b. Let the cipher text of the message of paragraph 4*b* be set down again, as in Figure 10*c*:

1	2	3	4	5	6	7	8	9	10	11	12	13
T	M	C	W	J	V	M	P	S	G	X	C	L
D	C	N	I	N	O	N	Y	G	U	O	I	N
P	E	T	X	Q	G	T	R	X	F	J	I	M
C	E	E	X	U	J	T	W	D	Y	X	A	Z
R	K	G	V	A	M	X	K	F	O	D	W	N
G	L	K	F	B	H	P	F	W	Q	Z	R	H
X	S	K	F	N	M	I	A	J	C	F	G	K
P	Y	X	I	Y	M	P	R	X	E	O	P	Q
W	W	R	V	C	W	J	S	E	W	F	Z	M
C	L	O	P	I	U	G	W	A	X	W	U	G
V	M	F	Y	X	J	X	W	Z	F	W	E	V
E	U	R	Z	R	H	H	G	U	T	Q	B	G

FIGURE 10c.

Using direct standard alphabets (Vigenère method), "decipher" the second line by means of the first line, that is, taking the letters of the second line as cipher text, those of the first line as key letters. Then use the thus-found "plain text" as "key letters" and "decipher" the third line of Figure 10*c*, as shown in Figure 14. Thus:

"Key"	T	M	C	W	J	V	M	P	S	G	X	C	L
Cipher	D	C	N	I	N	O	N	Y	G	U	O	I	N
"Plain"	K	Q	L	M	E	T	Z	J	O	O	R	G	C
"Key"	K	Q	L	M	E	T	Z	J	O	O	R	G	C
Cipher	P	E	T	X	Q	G	T	R	X	F	J	I	M
"Plain"	F	O	I	L	M	N	U	I	J	R	S	C	K

FIGURE 14.

Continue this operation for all the remaining lines of Figure 10*c* and write down the results in lines of 26 letters. Thus:

1	2	3	4	5	6	7	8	9	10	11	12	13	14	15	16	17	18	19	20	21	22	23	24	25	26
T	M	C	W	J	V	M	P	S	G	X	C	L	K	Q	L	M	E	T	Z	J	O	O	R	G	C
F	O	I	L	M	N	U	I	J	R	S	C	K	X	Q	W	M	I	W	Z	O	U	H	F	Y	P
U	U	K	J	S	Q	Y	W	L	H	Y	Y	Y	M	R	A	W	J	R	R	J	L	J	B	T	J
L	B	K	J	E	V	R	R	Y	T	E	N	B	E	X	N	Z	U	R	Y	A	Z	L	K	C	P
S	Z	E	W	I	F	L	S	F	L	V	X	X	K	M	K	T	A	P	V	E	V	M	B	X	J
L	A	V	F	X	U	C	S	E	T	V	H	M	T	U	W	U	U	N	F	O	Q	A	V	U	U

FIGURE 15.

Now write down the real plain text of the message in lines of 26 letters. Thus:

1	2	3	4	5	6	7	8	9	10	11	12	13	14	15	16	17	18	19	20	21	22	23	24	25	26
M	Y	L	E	F	T	F	L	A	N	K	I	S	R	E	C	E	I	V	I	N	G	H	E	A	V
Y	A	R	T	I	L	L	E	R	Y	F	I	R	E	E	N	E	M	Y	I	S	M	A	S	S	I
N	G	T	R	O	O	P	S	T	O	L	E	F	T	F	R	O	N	T	A	N	D	C	O	N	C
E	N	T	R	A	T	I	N	G	A	R	T	I	L	L	E	R	Y	T	H	E	R	E	X	W	I
L	L	N	E	E	D	C	O	N	S	I	D	E	R	A	B	L	E	R	E	I	N	F	O	R	C
E	M	E	N	T	S	T	O	M	A	I	N	T	A	I	N	M	Y	P	O	S	I	T	I	O	N

FIGURE 16.

c. When the underlined repetitions in Figures 15 and 16 are compared, they are found to be identical in the respective columns, and if the columns of Figure 15 are tested, they will be found to be monoalphabetic. The cipher message now gives every indication of being a repeating-key cipher. It is not difficult to explain this phenomenon in the light of the demonstration given in paragraph 3*g*. First, let the key word HORSECHESTNUT be enciphered by the following alphabet:

A	B	C	D	E	F	G	H	I	J	K	L	M	N	O	P	Q	R	S	T	U	V	W	X	Y	Z
A	Z	Y	X	W	V	U	T	S	R	Q	P	O	N	M	L	K	J	I	H	G	F	E	D	C	B

"Plain"	H	O	R	S	E	C	H	E	S	T	N	U	T
"Cipher"	T	M	J	I	W	Y	T	W	I	H	N	G	H

Then let the message MY LEFT FLANK, etc., be enciphered by direct standard alphabets as before, but for the key add the monoalphabetic equivalents of HORSECHESTNUT TMJIW. . . to the key itself, that is, use the 26-letter key HORSECHESTNUTTMJIWYTWIHNGH in a repeating-key manner. Thus (Fig. 17):

	1	2	3	4	5	6	7	8	9	10	11	12	13	14	15	16	17	18	19	20	21	22	23	24	25	26
Key	H	O	R	S	E	C	H	E	S	T	N	U	T	T	M	J	I	W	Y	T	W	I	H	N	G	H
Plain	M	Y	L	E	F	T	F	L	A	N	K	I	S	R	E	C	E	I	V	I	N	G	H	E	A	V
Cipher	T	M	C	W	J	V	M	P	S	G	X	C	L	K	Q	L	M	E	T	Z	J	O	O	R	G	C
Plain	Y	A	R	T	I	L	L	E	R	Y	F	I	R	E	E	N	E	M	Y	I	S	M	A	S	S	I
Cipher	F	O	I	L	M	N	U	I	J	R	S	C	K	X	Q	W	M	I	W	Z	O	U	H	F	Y	P
Plain	N	G	T	R	O	O	P	S	T	O	L	E	F	T	F	R	O	N	T	A	N	D	C	O	N	C
Cipher	U	U	K	J	S	Q	Y	W	L	H	Y	Y	Y	M	R	A	W	J	R	R	J	L	J	B	T	J
Plain	E	N	T	R	A	T	I	N	G	A	R	T	I	L	L	E	R	Y	T	H	E	R	E	X	W	I
Cipher	L	B	K	J	E	V	R	R	Y	T	E	N	B	E	X	N	Z	U	R	Y	A	Z	L	K	C	P
Plain	L	L	N	E	E	D	C	O	N	S	I	D	E	R	A	B	L	E	R	E	I	N	F	O	R	C
Cipher	S	Z	E	W	I	F	L	S	F	L	V	X	X	K	M	K	T	A	P	V	E	V	M	B	X	J
Plain	E	M	E	N	T	S	T	O	M	A	I	N	T	A	I	N	M	Y	P	O	S	I	T	I	O	N
Cipher	L	A	V	F	X	U	C	S	E	T	V	H	M	T	U	W	U	U	N	F	O	Q	A	V	U	U

FIGURE 17.

The cipher resultants of this process of *enciphering* a message coincide exactly with those obtained from the "*deciphering*" operation that gave rise to Figure 15. How does this happen?

462861 O - 42 - 9

d. First, let it be noted that the sequence TMJI . . . , which forms the second half of the key for enciphering the text in Figure 17 may be described as the standard alphabet *complement* of the sequence HORSECHESTNUT, which forms the first half of that key. Arithmetically, the sum of a letter of the first half and its homologous letter in the second half is 26. Thus:

$$
\begin{aligned}
H+T&= 7+19=26=0\\
O+M&=14+12=26=0\\
R+J&=17+ 9=26=0\\
S+I&=18+ 8=26=0\\
E+W&= 4+22=26=0
\end{aligned}
$$

That is, every letter of HORSECHESTNUT plus its homologous letter of the sequence TMJIWYTYIHNGH equals 26, which is here the same as zero. In other words, the sequence TMJIWYTWIHNGH is, by cryptographic arithmetic, equivalent to "minus HORSECHESTNUT." Therefore, in Figure 17, *enciphering* the second half of each line by the key letters TMJIWYTWIHNCH (i. e., adding 19, 12, 9, 8, . . .) is the same as *deciphering* by the key letters HORSECHESTNUT (i. e., subtracting 7, 14, 17, 18, . . .). For example:

$R_p(T_k)=17+19=36=10=K$, and
$R_p(-H_k)=17-7=10=K$

$E_p(M_k)=4+12=16=Q_c$, and
$E_p(-O_k)=4-14=(26+4)-14=16=Q_c$, and so on.

e. Refer now to Figure 15. The letters in the first half of line 1, beginning TMCWJ . . . are identical with those in the first half of line 1 of Figure 17. They must be identical because they are produced from identical elements. The letters in the second half of this same line in Figure 15, beginning KQLME . . . were produced by *deciphering* the letters in the second line of Figure 10*c*. Thus (taking for illustrative purposes only the first five letters in each case):

K Q L M E = D C N I N − T M C W J
But D C N I N = R E C E I + M Y L E F
And T M C W J = M Y L E F + H O R S E
Hence, K Q L M E = (R E C E I + M Y L E F) − (M Y L E F + H O R S E)
Or, K Q L M E = R E C E I − H O R S E (1)

As for the letters in the second half of line 1 of Figure 17, also beginning KQLME . . . , these letters were the result of *enciphering* RECEI by TMJIW.
Thus:

K Q L M E = R E C E I + T M J I W

But it has been shown in subparagraph *d* above that

T M J I W = − H O R S E

Hence, K Q L M E = R E C E I + (− H O R S E)
Or, K Q L M E = R E C E I − H O R S E (2)

Thus, equations (1) and (2) turn out to be identical but from what appear to be quite diverse sources.

f. What has been demonstrated in connection with the letters in line 1 of Figures 15 and 17 holds true for the letters in the other lines of these two figures, and it is not necessary to repeat the explanation. The steps show that the originally aperiodic, auto-key cipher has been converted, through a knowledge of the primary components, into a repeating-key cipher with a period twice the length of the introductory key. The message may now be solved as an ordinary repeating-key cipher.

g. (1) The foregoing case is based upon encipherment by the enciphering equations $\Theta_{k/2}=\Theta_{1/1}$; $\Theta_{p/1}=\Theta_{c/2}$. When encipherment by the enciphering equations $\Theta_{k/2}=\Theta_{1/1}$; $\Theta_{p/2}=\Theta_{c/1}$ has been followed, the conversion of a plain-text auto-keyed cipher yields a repeating-key cipher with a period equal to the length of the introductory key. In this conversion, the enciphering equations $\Theta_{k/2}=\Theta_{1/1}$; $\Theta_{p/1}=\Theta_{c/2}$ are used in finding equivalents.

(2) An example may be useful. Note the encipherment of the following message by auto-key method by enciphering equations $\Theta_{k/2}=\Theta_{1/1}$; $\Theta_{p/2}=\Theta_{c/1}$.

```
T U E S D A Y|I N F O R M A T I O N F R O M R E L I A B L E S O U R C E S I N D I C
I N F O R M A T I O N F R O M R E L I A B L E S O U R C E S I N D I C A T E S T H E
P T B W O M C L V J Z O F O T J Q Y D J N Z N O D M R B T O Q Z J R A W B W F Q Z C
```

(3) If the message is written out in lines corresponding to the length of the introductory key, and each line is *enciphered* by the one directly above it, using the enciphering equations $\Theta_{k/2}=\Theta_{1/1}$; $\Theta_{p/1}=\Theta_{c/2}$ in finding equivalents, the results are as shown in Figure 22*b*. But if the same message is *enciphered* by equations $\Theta_{k/2}=\Theta_{1/1}$; $\Theta_{p/2}=\Theta_{c/1}$, using the word TUESDAY as a repeating key, the cipher text (Fig. 18*c*) is identical with that obtained in Figure 18*b* by enciphering each successive line with the line above it.

Original cipher text		Original cipher text and converted text		Repeating key encipherment
				T U E S D A Y
				I N F O R M A
P T B W O M C	⟷	P T B W O M C	⟷	P T B W O M C
L V J Z O F O	⟷	L V J Z O F O		
				T I O N F R O
		A O K V C R Q	⟷	A O K V C R Q
T J Q Y D J N	⟷	T J Q Y D J N		
				M R E L I A B
		T X A T F A D	⟷	T X A T F A D
Z N O D M R B	⟷	Z N O D M R B		
				L E S O U R C
		S K O W R R E	⟷	S K O W R R E
T O Q Z J R A	⟷	T O Q Z J R A		
				E S I N D I C
		L Y E V A I E	⟷	L Y E V A I E
W B W F Q Z C	⟷	W B W F Q Z C		
				A T E S T H E
		H Z A A Q H G	⟷	H Z A A Q H G
a		*b*		*c*

FIGURE 18.

(4) Now note that the sequences joined by arrows in Figure 18 *b* and *c* are identical and since it is certain that Figure 18*c* is periodic in form because it was enciphered by the repeating-key method, it follows that Figure 18*b* is now also in periodic form, and in that form the message could be solved as though it were a repeating-key cipher.

h (1) In case of primary components consisting of a direct normal sequence sliding against a reversed normal (U. S. Army disk), the process of converting the auto-key text to periodic terms is accomplished by using two direct normal sequences and "deciphering" each line of the text (as transcribed in periods) by the line above it. For example, here is a message auto-enciphered by the aforementioned disk, with the initial key word TUESDAY:

```
T U E S D A Y|I N F O R M A T I O N F R O M R E L I A B L E S O U R C E S I N D I C
I N F O R M A T I O N F R O M R E L I A B L E S O U R C E S I N D I C A T E S T H E
L H Z E M O Y P F R B M V M H R K C X R N B N M X O J Z H M K B R J A E Z E V K B Y
```

(2) The cipher text is transcribed in periods equal to the length of the initial key word (7 letters) and the 2d line is "deciphered" with key letters of the 1st line, using enciphering equations $\Theta_{k/2}=\Theta_{1/1}$; $\Theta_{p/1}=\Theta_{c/2}$. The resultant letters are then used as key letters to "decipher" the 3d line of text and so on. The results are as seen in Figure 19*b*. Now let the original message be *enciphered* in repeating-key manner by the disk, with the key word TUESDAY, and the result is Figure 19*c*. Note that the odd or alternate lines of Figure 19*b* and *c* are identical, showing that the auto-key text has been converted into repeating-key text.

Original cipher text		Original cipher text and converted text		Repeating key encipherment
				T U E S D A Y
				I N F O R M A
L H Z E M O Y	←→	L H Z E M O Y	←→	L H Z E M O Y
P F R B M V M	←→	P F R B M V M		T I O N F R O
		A M Q F Y J K	←→	A M Q F Y J K
H R K C X R N	←→	H R K C X R N		M R E L I A B
		H D A H V A X	←→	H D A H V A X
B N M X O J Z	←→	B N M X O J Z		L E S O U R C
		I Q M E J J W	←→	I Q M E J J W
H M K B R J A	←→	H M K B R J A		E S I N D I C
		P C W F A S W	←→	P C W F A S W
E Z E V K B Y	←→	E Z E V K B Y		A T E S T H E
		T B A A K T U	←→	T B A A K T U
a		*b*		*c*

FIGURE 19.

i. The foregoing procedures indicate a simple method of solving ciphers of the foregoing types, when the primary components or the secondary cipher alphabets are known. It consists in assuming introductory keys of various lengths, converting the cipher text into repeating-key form, and then examining the resulting diagrams for repetitions. When a correct key length is assumed, repetitions will be as numerous as should be expected in ciphers of the repeating-key class; incorrect assumptions for key length will not show so many repetitions.

j. All the foregoing presupposes a knowledge of the cipher alphabets involved. When these are unknown, recourse must be had to first principles and the messages must be solved purely upon the basis of probable words, and repetitions, as outlined in paragraphs 27–28.

INDEX

Analytical Key for Military Cryptanalysis, Part III

[Numbers in parentheses refer to Paragraph Numbers in this text]

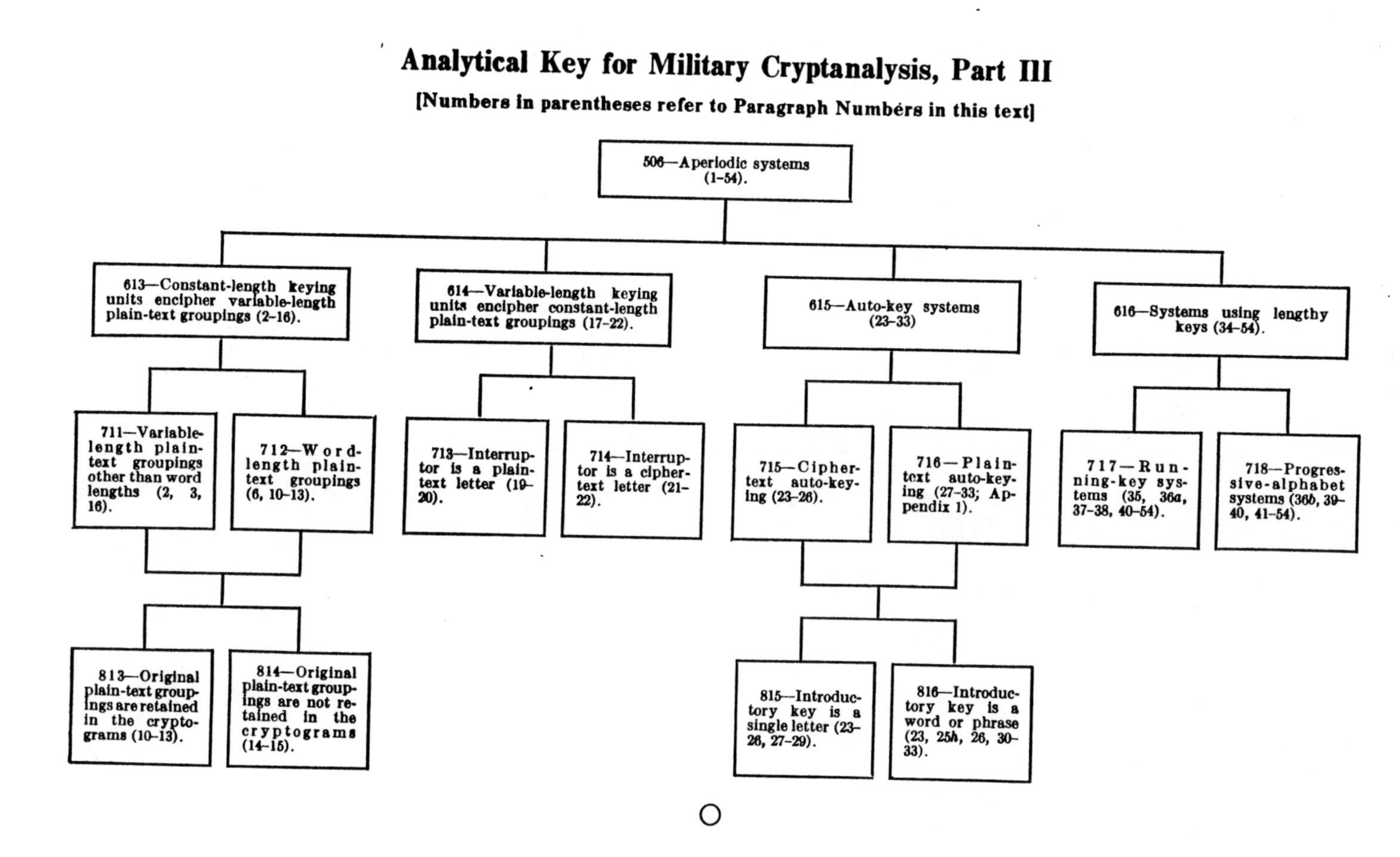